Molecular Biology

Molecular Biology

Vanessa Melton

SYRAWOOD
PUBLISHING HOUSE
New York

Published by Syrawood Publishing House,
750 Third Avenue, 9th Floor,
New York, NY 10017, USA
www.syrawoodpublishinghouse.com

Molecular Biology
Vanessa Melton

International Standard Book Number: 978-1-64740-095-8 (Hardback)

Cataloging-in-Publication Data

Molecular biology / Vanessa Melton.
    p. cm.
Includes bibliographical references and index.
ISBN 978-1-64740-095-8
1. Molecular biology. 2. Biomolecules.
3. Molecular biology--Technique.
I. Melton, Vanessa.
QH506 .M65 2022
572.8--dc23

TABLE OF CONTENTS

Preface VII

Chapter 1   **What is Molecular Biology?** 1

     • Cell   1

Chapter 2   **Biomolecules** 36

     • Protein   37

     • Nucleic Acid   51

Chapter 3   **Gene Expression and Regulation** 60

     • Gene expression   60

     • Central Dogma of Molecular Biology   61

     • DNA Replication   66

     • Transcription   77

     • Translation   84

Chapter 4   **Molecular Biology: Techniques and Processes** 91

     • Gel Electrophoresis   91

     • Restriction Digest   93

     • Polymerase Chain Reaction   95

     • Ligation   98

     • Blot   103

     • DNA Sequencing   105

     • Molecular Cloning   149

     • DNA Cloning   151

     • DNA Microarray   154

     • Cell Culture   163

     • Transfection   173

Chapter 5   **Diverse Aspects in Molecular Biology** 178

     • DNA Replication   178

     • Genome   182

- Proteome   188
- Protein Biosynthesis   189
- Oligonucleotide Synthesis   193
- Post-transcriptional Modification   206
- Post-translational Modification   208
- Recombinant DNA   213

Permissions

Index

This book aims to help a broader range of students by exploring a wide variety of significant topics related to this discipline. It will help students in achieving a higher level of understanding of the subject and excel in their respective fields. This book would not have been possible without the unwavered support of my senior professors who took out the time to provide me feedback and help me with the process. I would also like to thank my family for their patience and support.

Molecular biology is a biological branch that is concerned with the molecular aspects of biological activity that occurs between biomolecules within a cell. It observes and regulates the interactions between proteins, DNA, RNA and their biosynthesis. In molecular biology specific techniques are combined with the techniques and concepts from genetics and biochemistry. It is the study of molecular basis of various processes including replication, translation, transcription and cell function. Some of the common techniques of this domain include molecular cloning, gel electrophoresis, macromolecule blotting. This book presents the complex subject of molecular biology in the most comprehensible and easy to understand language. The topics covered in this extensive book deal with the core subjects of this biological field. This book will prove to be immensely beneficial to students and experts in this field.

A brief overview of the book contents is provided below:

Chapter – What is Molecular Biology?

The branch of biology that is concerned with the study of composition, structure and interactions of biomolecules of a cell is known as molecular biology. It focuses on the interactions and biosynthesis of proteins, RNA and DNA. This is an introductory chapter which will briefly introduce all the significant aspects of molecular biology.

Chapter – Biomolecules

Biomolecules are the macromolecules and ions that are present in organisms. Macromolecules such as lipids, nucleic acid, proteins, etc. are essential for various biological processes such as cell morphogenesis, cell division and development. This chapter has been carefully written to provide an easy understanding of these biomolecules.

Chapter – Gene Expression and Regulation

Gene expression is the process through which the genetic code of a gene is used to synthesize the functional gene product. Gene regulation includes the mechanisms that are used by cells to alter the structural and chemical changes in a gene and DNA elements for regulating transcription. This chapter discusses in detail the concepts and processes related to gene expression and regulation.

Chapter – Molecular Biology: Techniques and Processes

Many important techniques are used in molecular biology to understand various biological processes. Some of these are DNA sequencing, molecular cloning, gel electrophoresis, polymerase chain reaction, DNA microarray, transfection and ligation. The topics elaborated in this chapter will help in gaining a better perspective about these techniques used in molecular biology.

Chapter – Diverse Aspects in Molecular Biology

Some of the fundamental concepts studied in molecular biology include DNA and RNA replication, protein biosynthesis, oligonucleotide synthesis, genome and proteome. This chapter closely examines these fundamental concepts of molecular biology to provide an extensive understanding of the subject.

Vanessa Melton

CHAPTER 1

What is Molecular Biology?

The branch of biology that is concerned with the study of composition, structure and interactions of biomolecules of a cell is known as molecular biology. It focuses on the interactions and biosynthesis of proteins, RNA and DNA. This is an introductory chapter which will briefly introduce all the significant aspects of molecular biology.

Molecular biology is the field of science concerned with studying the chemical structures and processes of biological phenomena that involve the basic units of life, molecules. Of growing importance since the 1940s, molecular biology developed out of the related fields of biochemistry, genetics, and biophysics. The discipline is particularly concerned with the study of proteins and nucleic acids—i.e., the macromolecules that are essential to life processes. Molecular biology seeks to understand the three-dimensional structure of these macromolecules through such techniques as X-ray diffraction and electron microscopy. The discipline particularly seeks to understand the molecular basis of genetic processes; molecular biologists map the location of genes on specific chromosomes, associate these genes with particular characters of an organism, and use recombinant DNA technology to isolate, sequence, and modify specific genes.

In its early period during the 1940s, the field was concerned with elucidating the basic three-dimensional structure of proteins. Growing knowledge of the structure of proteins in the early 1950s enabled the structure of deoxyribonucleic acid (DNA)—the genetic blueprint found in all living things—to be described in 1953. Further research enabled scientists to gain an increasingly detailed knowledge not only of DNA and ribonucleic acid (RNA) but also of the chemical sequences within these substances that instruct the cells and viruses to make proteins.

Molecular biology remained a pure science with few practical applications until the 1970s, when certain types of enzymes were discovered that could cut and recombine segments of DNA in the chromosomes of certain bacteria. The resulting recombinant DNA technology became one of the most active branches of molecular biology because it allows the manipulation of the genetic sequences that determine the basic characters of organisms.

CELL

Cell is the basic membrane-bound unit that contains the fundamental molecules of life and of which all living things are composed. A single cell is often a complete organism in itself, such as a bacterium or yeast. Other cells acquire specialized functions as they mature. These cells cooperate with other specialized cells and become the building blocks of large multicellular organisms, such as humans and other animals. Although cells are much larger than atoms, they are still very small. The smallest known cells are a group of tiny bacteria called mycoplasmas; some of these single-celled organisms are spheres as small as 0.2 µm in diameter (1µm = about 0.000039 inch), with a total mass of 10–14 gram—equal to that of 8,000,000,000 hydrogen atoms. Cells of humans typically have a mass 400,000 times larger than the mass of a single mycoplasma bacterium, but even human cells are only about 20 µm across. It would require a sheet of about 10,000 human cells to cover the head of a pin, and each human organism is composed of more than 75,000,000,000,000 cells.

Nature and Function of Cells

A cell is enclosed by a plasma membrane, which forms a selective barrier that allows nutrients to enter and waste products to leave. The interior of the cell is organized into many specialized compartments, or organelles, each surrounded by a separate membrane. One major organelle, the nucleus, contains the genetic information necessary for cell growth and reproduction. Each cell contains only one nucleus, whereas other types of organelles are present in multiple copies in the cellular contents, or cytoplasm. Organelles include mitochondria, which are responsible for the energy transactions necessary for cell survival; lysosomes, which digest unwanted materials within the cell; and the endoplasmic reticulum and the Golgi apparatus, which play important roles in the internal organization of the cell by synthesizing selected molecules and then processing, sorting, and directing them to their proper locations. In addition, plant cells contain chloroplasts, which are responsible for photosynthesis, whereby the energy of sunlight is used to convert molecules of carbon dioxide (CO_2) and water (H_2O) into carbohydrates. Between all these organelles is the space in the cytoplasm called the cytosol. The cytosol contains an organized framework of fibrous molecules that constitute the cytoskeleton, which gives a cell its shape, enables organelles to move within the cell, and provides a mechanism by which the cell itself can move. The cytosol also contains more than 10,000 different kinds of molecules that are involved in cellular biosynthesis, the process of making large biological molecules from small ones.

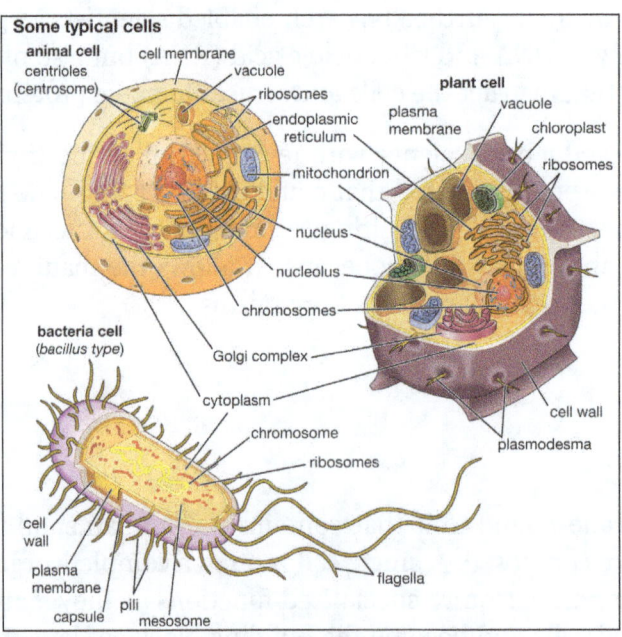

Animal cells and plant cells contain membrane-bound organelles, including a distinct nucleus. In contrast, bacterial cells do not contain organelles.

Specialized organelles are a characteristic of cells of organisms known as eukaryotes. In contrast, cells of organisms known as prokaryotes do not contain organelles and are generally smaller than eukaryotic cells. However, all cells share strong similarities in biochemical function.

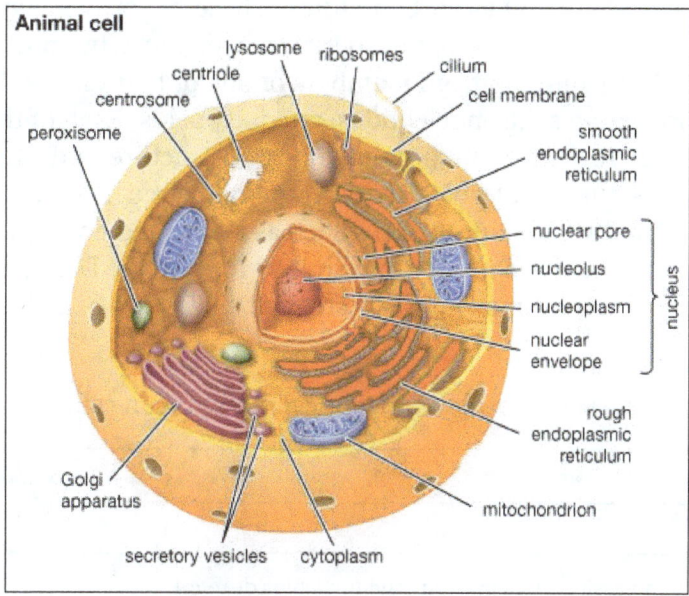

Cutaway drawing of a eukaryotic cell.

Molecules of Cells

Cells contain a special collection of molecules that are enclosed by a membrane. These molecules give cells the ability to grow and reproduce. The overall process of cellular reproduction occurs in two steps: cell growth and cell division. During cell growth, the cell ingests certain molecules from its surroundings by selectively carrying them through its cell membrane. Once inside the cell, these molecules are subjected to the action of highly specialized, large, elaborately folded molecules called enzymes. Enzymes act as catalysts by binding to ingested molecules and regulating the rate at which they are chemically altered. These chemical alterations make the molecules more useful to the cell. Unlike the ingested molecules, catalysts are not chemically altered themselves during the reaction, allowing one catalyst to regulate a specific chemical reaction in many molecules.

Biological catalysts create chains of reactions. In other words, a molecule chemically transformed by one catalyst serves as the starting material, or substrate, of a second catalyst and so on. In this way, catalysts use the small molecules brought into the cell from the outside environment to create increasingly complex reaction products. These products are used for cell growth and the replication of genetic material. Once the genetic material has been copied and there are sufficient molecules to support cell division, the cell divides to create two daughter cells. Through many such cycles of cell growth and division, each parent cell can give rise to millions of daughter cells, in the process converting large amounts of inanimate matter into biologically active molecules.

Structure of Biological Molecules

Cells are largely composed of compounds that contain carbon. The study of how carbon atoms interact with other atoms in molecular compounds forms the basis of the field of organic chemistry and plays a large role in understanding the basic functions of cells. Because carbon atoms can form stable bonds with four other atoms, they are uniquely suited for the construction of complex molecules. These complex molecules are typically made up of chains and rings that contain hydrogen, oxygen, and nitrogen atoms, as well as carbon atoms. These molecules may consist of anywhere from 10 to millions of atoms linked together in specific arrays. Most, but not all, of the carbon-containing molecules in cells are built up from members of one of four different families of small organic molecules: sugars, amino acids, nucleotides, and fatty acids. Each of these families contains a group of molecules that resemble one another in both structure and function. In addition to other important functions, these molecules are used to build large macromolecules. For example, the sugars can be linked to form polysaccharides such as starch and glycogen, the amino acids can be linked to form proteins, the nucleotides can be linked to form the DNA (deoxyribonucleic acid) and RNA (ribonucleic acid) of chromosomes, and the fatty acids can be linked to form the lipids of all cell membranes.

Table: Approximate chemical composition of a typical mammalian cell.

Component	Percent of total cell weight
Water	70
Inorganic ions (sodium, potassium, magnesium, calcium, chloride, etc.)	1
Miscellaneous small metabolites	3
Proteins	18
RNA	1.1
DNA	0.25
Phospholipids and other lipids	5
Polysaccharides	2

Aside from water, which forms 70 percent of a cell's mass, a cell is composed mostly of macromolecules. By far the largest portion of macromolecules are the proteins. An average-sized protein macromolecule contains a string of about 400 amino acid molecules. Each amino acid has a different side chain of atoms that interact with the atoms of side chains of other amino acids. These interactions are very specific and cause the entire protein molecule to fold into a compact globular form. In theory, nearly an infinite variety of proteins can be formed, each with a different sequence of amino acids. However, nearly all these proteins would fail to fold in the unique ways required to form efficient functional surfaces and would therefore be useless to the cell. The proteins present in cells of modern animals and humans are products of a long evolutionary history, during which the ancestor proteins were naturally selected for their ability to fold into specific three-dimensional forms with unique functional surfaces useful for cell survival.

Most of the catalytic macromolecules in cells are enzymes. The majority of enzymes are proteins. Key to the catalytic property of an enzyme is its tendency to undergo a change in its shape when it binds to its substrate, thus bringing together reactive groups on substrate molecules. Some enzymes are macromolecules of RNA, called ribozymes. Ribozymes consist of linear chains of nucleotides that fold in specific ways to form unique surfaces, similar to the ways in which proteins fold.

As with proteins, the specific sequence of nucleotide subunits in an RNA chain gives each macro-molecule a unique character. RNA molecules are much less frequently used as catalysts in cells than are protein molecules, presumably because proteins, with the greater variety of amino acid side chains, are more diverse and capable of complex shape changes. However, RNA molecules are thought to have preceded protein molecules during evolution and to have catalyzed most of the chemical reactions required before cells could evolve.

Coupled Chemical Reactions

Cells must obey the laws of chemistry and thermodynamics. When two molecules react with each other inside a cell, their atoms are rearranged, forming different molecules as reaction products and releasing or consuming energy in the process. Overall, chemical reactions occur only in one direction; that is, the final reaction product molecules cannot spontaneously react, in a reversal of the original process, to reform the original molecules. This directionality of chemical reactions is explained by the fact that molecules only change from states of higher free energy to states of lower free energy. Free energy is the ability to perform work (in this case, the "work" is the rearrange-ment of atoms in the chemical reaction). When work is performed, some free energy is used and lost, with the result that the process ends at lower free energy. To use a familiar mechanical anal-ogy, water at the top of a hill has the ability to perform the "work" of flowing downhill (i.e., it has high free energy), but, once it has flowed downhill, it cannot flow back up (i.e., it is in a state of low free energy). However, through another work process—that of a pump, for example—the water can be returned to the top of the hill, thereby recovering its ability to flow downhill. In thermodynamic terms, the free energy of the water has been increased by energy from an outside source (i.e., the pump). In the same way, the product molecules of a chemical reaction in a cell cannot reverse the reaction and return to their original state unless energy is supplied by coupling the process to an-other chemical reaction.

All catalysts, including enzymes, accelerate chemical reactions without affecting their direction. To return to the mechanical analogy, enzymes cannot make water flow uphill, although they can provide specific pathways for a downhill flow. Yet most of the chemical reactions that the cell needs to synthesize new molecules necessary for its growth require an uphill flow. In other words, the reactions require more energy than their starting molecules can provide.

Cells use a single strategy over and over again in order to get around the limitations of chem-istry: they use the energy from an energy-releasing chemical reaction to drive an energy-ab-sorbing reaction that would otherwise not occur. A useful mechanical analogy might be a mill wheel driven by the water in a stream. The water, in order to flow downhill, is forced to flow past the blades of the wheel, causing the wheel to turn. In this way, part of the energy from the moving stream is harnessed to move a mill wheel, which may be linked to a winch. As the winch turns, it can be used to pull a heavy load uphill. Thus, the energy-absorbing (but useful) uphill movement of a load can be driven by coupling it directly to the energy-releasing flow of water.

In cells, enzymes play the role of mill wheels by coupling energy-releasing reactions with ener-gy-absorbing reactions. As discussed below, in cells the most important energy-releasing reaction serving a role similar to that of the flowing stream is the hydrolysis of adenosine triphosphate (ATP). In turn, the production of ATP molecules in the cells is an energy-absorbing reaction that is driven by being coupled to the energy-releasing breakdown of sugar molecules. In retracing this chain of reactions, it is necessary first to understand the source of the sugar molecules.

Photosynthesis: The Beginning of the Food Chain

Sugar molecules are produced by the process of photosynthesis in plants and certain bacteria. These organisms lie at the base of the food chain, in that animals and other nonphotosynthesizing organisms depend on them for a constant supply of life-supporting organic molecules. Humans, for example, obtain these molecules by eating plants or other organisms that have previously eaten food derived from photosynthesizing organisms.

Plants and photosynthetic bacteria are unique in their ability to convert the freely available electromagnetic energy in sunlight into chemical bond energy, the energy that holds atoms together in molecules and is transferred or released in chemical reactions. The process of photosynthesis can be summarized by the following equation:

$$\text{Energy (solar)} + CO_2 + H_2O \rightarrow \text{Sugar molecules} + O_2.$$

The energy-absorbing photosynthetic reaction is the reverse of the energy-releasing oxidative decomposition of sugar molecules. During photosynthesis, chlorophyll molecules absorb energy from sunlight and use it to fuel the production of simple sugars and other carbohydrates. The resulting abundance of sugar molecules and related biological products makes possible the existence of nonphotosynthesizing life on Earth.

ATP: Fueling Chemical Reactions

Certain enzymes catalyze the breakdown of organic foodstuffs. Once sugars are transported into cells, they either serve as building blocks in the form of amino acids for proteins and fatty acids for lipids or are subjected to metabolic pathways to provide the cell with ATP. ATP, the common carrier of energy inside the cell, is made from adenosine diphosphate (ADP) and inorganic phosphate (P_i). Stored in the chemical bond holding the terminal phosphate compound onto the ATP molecule is the energy derived from the breakdown of sugars. The removal of the terminal phosphate, through the water-mediated reaction called hydrolysis, releases this energy, which in turn fuels a large number of crucial energy-absorbing reactions in the cell. Hydrolysis can be summarized as follows:

$$ATP + H_2O \rightarrow ADP + P_i + \text{Energy}.$$

The formation of ATP is the reverse of this equation, requiring the addition of energy. The central cellular pathway of ATP synthesis begins with glycolysis, a form of fermentation in which the sugar glucose is transformed into other sugars in a series of nine enzymatic reactions, each successive reaction involving an intermediate sugar containing phosphate. In the process, the six-carbon glucose is converted into two molecules of the three-carbon pyruvic acid. Some of the energy released through glycolysis of each glucose molecule is captured in the formation of two ATP molecules.

The second stage in the metabolism of sugars is a set of interrelated reactions called the tricarboxylic acid cycle. This cycle takes the three-carbon pyruvic acid produced in glycolysis and uses its carbon atoms to form carbon dioxide (CO_2) while transferring its hydrogen atoms to special carrier molecules, where they are held in high-energy linkage.

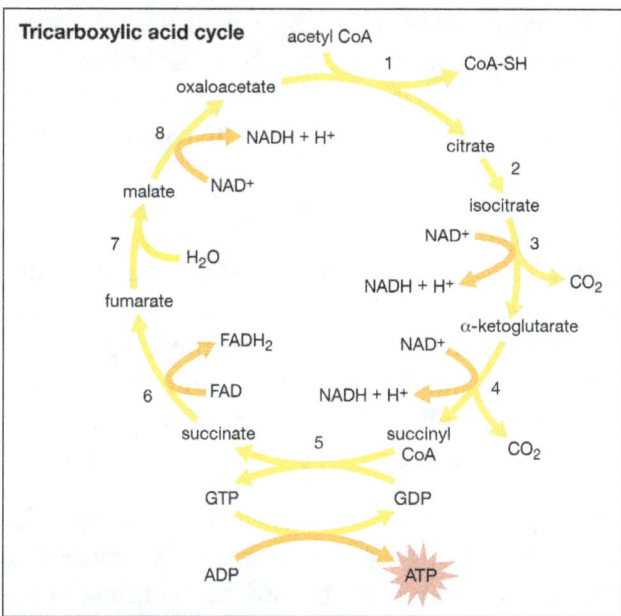

The eight-step tricarboxylic acid cycle.

In the third and last stage in the breakdown of sugars, oxidative phosphorylation, the high-energy hydrogen atoms are first separated into protons and high-energy electrons. The electrons are then passed from one electron carrier to another by means of an electron-transport chain. Each electron carrier in the chain has an increasing affinity for electrons, with the final electron acceptor being molecular oxygen (O_2). As separated electrons and protons, the hydrogen atoms are transferred to O_2 to form water. This reaction releases a large amount of energy, which drives the synthesis of a large number of ATP molecules from ADP and Pi.

Most of the cell's ATP is produced when the products of glycolysis are oxidized completely by a combination of the tricarboxylic acid cycle and oxidative phosphorylation. The process of glycolysis alone produces relatively small amounts of ATP. Glycolysis is an anaerobic reaction; that is, it can occur even in the absence of oxygen. The tricarboxylic acid cycle and oxidative phosphorylation, on the other hand, require oxygen. Glycolysis forms the basis of anaerobic fermentation, and it presumably was a major source of ATP for early life on Earth, when very little oxygen was available in the atmosphere. Eventually, however, bacteria evolved that were able to carry out photosynthesis. Photosynthesis liberated these bacteria from a dependence on the metabolism of organic materials that had accumulated from natural processes, and it also released oxygen into the atmosphere. Over a prolonged period of time, the concentration of molecular oxygen increased until it became freely available in the atmosphere. The aerobic tricarboxylic acid cycle and oxidative phosphorylation then evolved, and the resulting aerobic cells made much more efficient use of foodstuffs than their anaerobic ancestors, because they could convert much larger amounts of chemical bond energy into ATP.

Genetic Information of Cells

Cells can thus be seen as a self-replicating network of catalytic macromolecules engaged in a carefully balanced series of energy conversions that drive biosynthesis and cell movement. But energy alone is not enough to make self-reproduction possible; the cell must contain detailed instructions that dictate exactly how that energy is to be used. These instructions are analogous to the blueprints that a builder uses to construct a house; in the case of cells, however, the blueprints themselves must be duplicated along with the cell before it divides, so that each daughter cell can retain the instructions that it needs for its own replication. These instructions constitute the cell's heredity.

DNA: The Genetic Material

During the early 19th century, it became widely accepted that all living organisms are composed of cells arising only from the growth and division of other cells. The improvement of the microscope then led to an era during which many biologists made intensive observations of the microscopic structure of cells. By 1885, a substantial amount of indirect evidence indicated that chromosomes—dark-staining threads in the cell nucleus—carried the information for cell heredity. It was later shown that chromosomes are about half DNA and half protein by weight.

The revolutionary discovery suggesting that DNA molecules could provide the information for their own replication came in 1953, when American geneticist and biophysicist James Watson and British biophysicist Francis Crick proposed a model for the structure of the double-stranded DNA molecule (called the DNA double helix). In this model, each strand serves as a template in the synthesis of a complementary strand. Subsequent research confirmed the Watson and Crick model of DNA replication and showed that DNA carries the genetic information for reproduction of the entire cell.

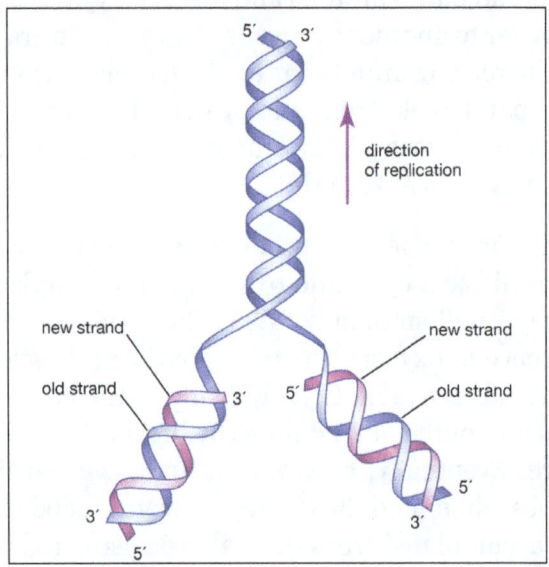

The initial proposal of the structure of DNA by James Watson and Francis Crick was accompanied by a suggestion on the means of replication.

All of the genetic information in a cell was initially thought to be confined to the DNA in the chromosomes of the cell nucleus. Later discoveries identified small amounts of additional genetic information present in the DNA of much smaller chromosomes located in two types of organelles in the cytoplasm. These organelles are the mitochondria in animal cells and the mitochondria and chloroplasts in plant cells. The special chromosomes carry the information coding for a few of the many proteins and RNA molecules needed by the organelles. They also hint at the evolutionary origin of these organelles, which are thought to have originated as free-living bacteria that were taken up by other organisms in the process of symbiosis.

RNA: Replicated from DNA

It is possible for RNA to replicate itself by mechanisms related to those used by DNA, even though it has a single-stranded instead of a double-stranded structure. In early cells RNA is thought to have replicated itself in this way. However, all of the RNA in present-day cells is synthesized by special enzymes that construct a single-stranded RNA chain by using one strand of the DNA helix as a template. Although RNA molecules are synthesized in the cell nucleus, where the DNA is located, most of them are transported to the cytoplasm before they carry out their functions.

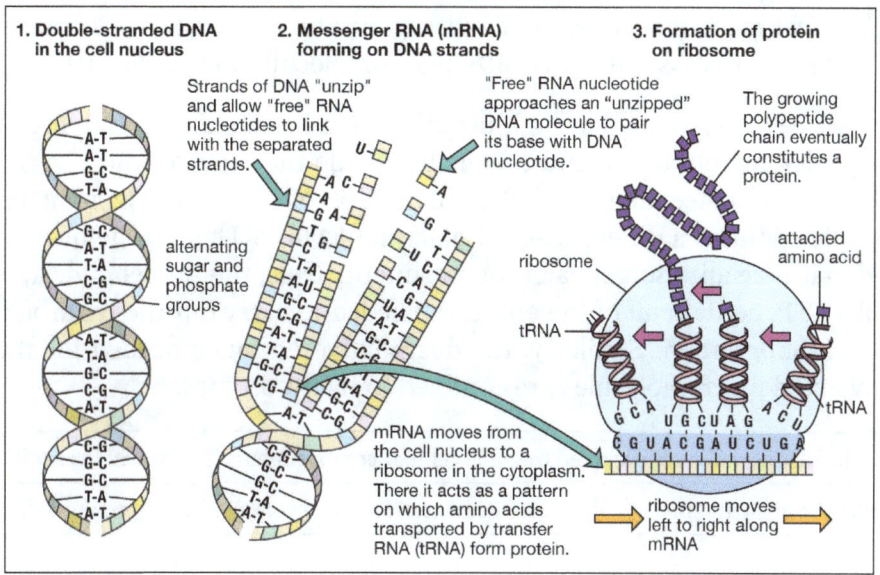

Molecular genetics emerged from the realization that DNA and RNA constitute the genetic material of all living organisms. (1) DNA, located in the cell nucleus, is made up of nucleotides that contain the bases adenine (A), thymine (T), guanine (G), and cytosine (C). (2) RNA, which contains uracil (U) instead of thymine, transports the genetic code to protein-synthesizing sites in the cell. (3) Messenger RNA (mRNA) then carries the genetic information to ribosomes in the cell cytoplasm that translate the genetic information into molecules of protein.

The RNA molecules in cells have two main roles. Some, the ribozymes, fold up in ways that allow them to serve as catalysts for specific chemical reactions. Others serve as "messenger RNA," which provides templates specifying the synthesis of proteins. Ribosomes, tiny protein-synthesizing machines located in the cytoplasm, "read" the messenger RNA molecules and "translate" them into proteins by using the genetic code. In this translation, the sequence of nucleotides in the messenger RNA chain is decoded three nucleotides at a time, and each nucleotide triplet (called a codon) specifies a particular amino acid. Thus, a nucleotide sequence in the DNA specifies a protein provided that a messenger RNA molecule is produced from that DNA sequence. Each region of the DNA sequence specifying a protein in this way is called a gene.

DNA molecules catalyze not only their own duplication but also dictate the structures of all protein molecules. A single human cell contains about 10,000 different proteins produced by the expression of 10,000 different genes. Actually, a set of human chromosomes is thought to contain DNA with enough information to express between 30,000 and 100,000 proteins, but most of these proteins seem to be made only in specialized types of cells and are therefore not present throughout the body.

Organization of Cells

Intracellular Communication

A cell with its many different DNA, RNA, and protein molecules is quite different from a test tube containing the same components. When a cell is dissolved in a test tube, thousands of different types of molecules randomly mix together. In the living cell, however, these components are kept in specific places, reflecting the high degree of organization essential for the growth and division of the cell. Maintaining this internal organization requires a continuous input of energy, because spontaneous chemical reactions always create disorganization. Thus, much of the energy released by ATP hydrolysis fuels processes that organize macromolecules inside the cell.

When a eukaryotic cell is examined at high magnification in an electron microscope, it becomes apparent that specific membrane-bound organelles divide the interior into a variety of subcompartments. Although not detectable in the electron microscope, it is clear from biochemical assays that each organelle contains a different set of macromolecules. This biochemical segregation reflects the functional specialization of each compartment. Thus, the mitochondria, which produce most of the cell's ATP, contain all of the enzymes needed to carry out the tricarboxylic acid cycle and oxidative phosphorylation. Similarly, the degradative enzymes needed for the intracellular digestion of unwanted macromolecules are confined to the lysosomes.

The relative volumes occupied by some cellular compartments in a typical liver cell		
Cellular compartment	Percent of total cell volume	Approximate number per cell
Cytosol	54	1
Mitochondrion	22	1,700
Endoplasmic reticulum plus Golgi apparatus	15	1
Nucleus	6	1
Lysosome	1	300

It is clear from this functional segregation that the many different proteins specified by the genes in the cell nucleus must be transported to the compartment where they will be used. Not surprisingly, the cell contains an extensive membrane-bound system devoted to maintaining just this intracellular order. The system serves as a post office, guaranteeing the proper routing of newly synthesized macromolecules to their proper destinations.

All proteins are synthesized on ribosomes located in the cytosol. As soon as the first portion of the amino acid sequence of a protein emerges from the ribosome, it is inspected for the presence of a short "endoplasmic reticulum (ER) signal sequence." Those ribosomes making proteins with such a sequence are transported to the surface of the ER membrane, where they complete their synthesis; the proteins made on these ribosomes are immediately transferred through the ER membrane to the inside of the ER compartment. Proteins lacking the ER signal sequence remain in the cytosol and are released from the ribosomes when their synthesis is completed. This chemical decision process places some newly completed protein chains in the cytosol and others within an extensive membrane-bounded compartment in the cytoplasm, representing the first step in intracellular protein sorting.

The newly made proteins in both cell compartments are then sorted further according to additional signal sequences that they contain. Some of the proteins in the cytosol remain there, while others go to the surface of mitochondria or (in plant cells) chloroplasts, where they are transferred through the membranes into the organelles. Subsignals on each of these proteins then designate exactly where in the organelle the protein belongs. The proteins initially sorted into the ER have an even wider range of destinations. Some of them remain in the ER, where they function as part of the organelle. Most enter transport vesicles and pass to the Golgi apparatus, separate membrane-bounded organelles that contain at least three subcompartments. Some of the proteins are retained in the subcompartments of the Golgi, where they are utilized for functions peculiar to that organelle. Most eventually enter vesicles that leave the Golgi for other cellular destinations such as the cell membrane, lysosomes, or special secretory vesicles.

Intercellular Communication

Formation of a multicellular organism starts with a small collection of similar cells in an embryo and proceeds by continuous cell division and specialization to produce an entire community of cooperating cells, each with its own role in the life of the organism. Through cell cooperation, the organism becomes much more than the sum of its component parts.

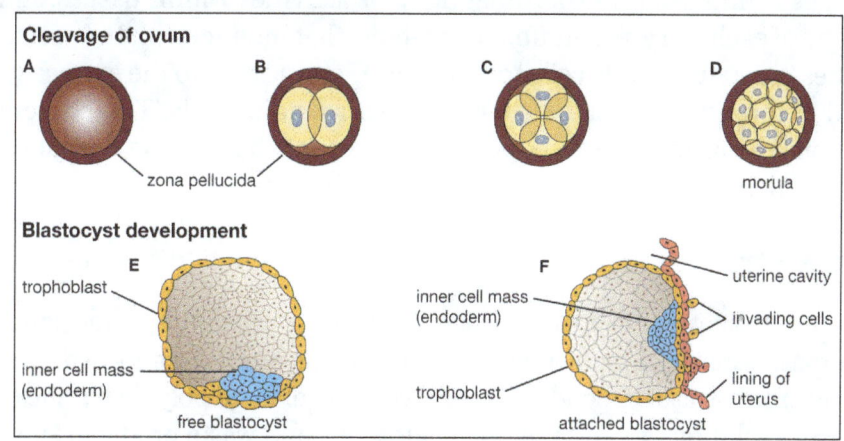

The ovum contains a small collection of cells in the early stages of human development.

As cells divide (A–D), they are separated into different regions of the ovum. Each region of the ovum transmits a unique set of chemical signals to nearby cells. Thus, the signals detected by one cell differ from those detected by its neighbour cells. In this process, known as cell determination, cells are individually programmed to direct them toward development into different cell types.

A fertilized egg multiplies and produces a whole family of daughter cells, each of which adopts a structure and function according to its position in the entire assembly. All of the daughter cells contain the same chromosomes and therefore the same genetic information. Despite this common inheritance, different types of cells behave differently and have different structures. In order for this to be the case, they must express different sets of genes, so that they produce different proteins despite their identical embryological ancestors.

During the development of an embryo, it is not sufficient for all the cell types found in the fully developed individual simply to be created. Each cell type must form in the right place at the right time and in the correct proportion; otherwise, there would be a jumble of randomly assorted cells in no way resembling an organism. The orderly development of an organism depends on a process called cell determination, in which initially identical cells become committed to different pathways of development. A fundamental part of cell determination is the ability of cells to detect different chemicals within different regions of the embryo. The chemical signals detected by one cell may be different from the signals detected by its neighbour cells. The signals that a cell detects activate a set of genes that tell the cell to differentiate in ways appropriate for its position within the embryo. The set of genes activated in one cell differs from the set of genes activated in the cells around it. The process of cell determination requires an elaborate system of cell-to-cell communication in early embryos.

Cell Matrix and Cell-to-Cell Communication

The development of single cells into multicellular organisms involves a number of adaptations. The cells become specialized, acquiring distinct functions that contribute to the survival of the organism. The behaviour of individual cells is also integrated with that of similar cells, so that they act together in a regulated fashion. To achieve this integration, cells assemble into specialized tissues, each tissue being composed of cells and the spaces outside of the cells.

The surface of cells is important in coordinating their activities within tissues. Embedded in the plasma membrane of each cell are a number of proteins that interact with the surface or secretions of other cells. These proteins enable cells to "recognize" and adhere to the extracellular matrix and one another and to form populations distinct from surrounding cells. These interactions are key to the organizational behaviour of cell populations and contribute to the formation of embryonic tissues and the function of normal tissue in the adult organism.

Extracellular Matrix

A substantial part of tissues is the space outside of the cells, called the extracellular space. This is filled with a composite material, known as the extracellular matrix, composed of a gel in which a number of fibrous proteins are suspended. The gel consists of large polysaccharide (complex sugar) molecules in a water solution of inorganic salts, nutrients, and waste products known as the interstitial fluid. The major types of protein in the matrix are structural proteins and adhesive proteins.

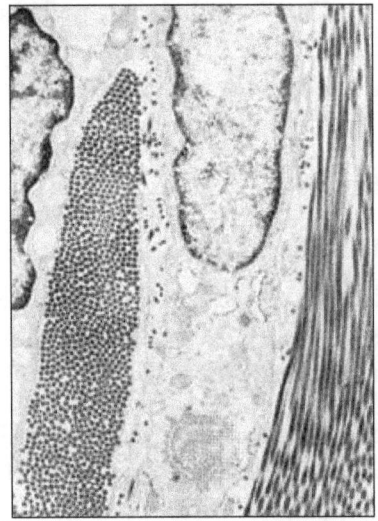

Electron micrograph of a small area of dense fibrous connective tissue, illustrating the intimate association of cells and fibres. In the centre is a portion of a fibrocyte, and on either side are two collagen fibres. The collagen fibre on the left is cut transversely, showing round cross sections of the unit fibrils. The collagen fibre on the right has been cut nearly parallel to its long axis and shows extensive segments of the cross-striated fibrils.

There are two general types of tissues distinct not only in their cellular organization but also in the composition of their extracellular matrix. The first type, mesenchymal tissue, is made up of clusters of cells grouped together but not closely adherent to one another. They synthesize a highly hydrated gel, rich in salts, fluid, and fibres, known as the interstitial matrix. Connective tissue is a mesenchyme that fastens together other more highly organized tissues. The solidity of various connective tissues varies according to the consistency of their extracellular matrix, which in turn depends on the water content of the gels, the amount and type of polysaccharides and structural proteins, and the presence of other salts. For example, bone is rich in calcium phosphate, giving that tissue its rigidity; tendons are mostly fibrous structural proteins, yielding a ropelike consistency; and joint spaces are filled with a lubricating fluid of mostly polysaccharide and interstitial fluid.

Epithelial tissues, the second type, are sheets of cells adhering at their side, or lateral, surfaces. They synthesize and deposit at their bottom, or basal, surfaces an organized complex of matrix materials known as the basal lamina or basement membrane. This thin layer serves as a boundary with connective tissue and as a substrate to which epithelial cells are attached.

Matrix Polysaccharides

The polysaccharides, or glycans, of the extracellular matrix are responsible for its gel-like quality and for organizing its components. These large acidic molecules exist alone (as glycosaminoglycans) or in combination with small proteins (as proteoglycans). They bind an extraordinarily large amount of water, thus forming massively swollen gels that fill the spaces between cells. Bound to proteins, they also organize other molecules in the extracellular matrix. The firmness and resiliency of cartilage, as at the surface of joints, is due to highly organized proteoglycans that bind water tightly.

Matrix Proteins

Matrix proteins are large molecules tightly bound to form extensive networks of insoluble fibres. These fibres may even exceed the size of the cells themselves. The proteins are of two general types, structural and adhesive.

The structural proteins, collagen and elastin, are the dominant matrix proteins. At least 10 different types of collagen are present in various tissues. The most common, type I collagen, is the most abundant protein in vertebrate animals, accounting for nearly 25 percent of the total protein in the body. The various collagen types share structural features, all being composed of three intertwined polypeptide chains. In some collagens the chains are linked together by covalent bonds, yielding a ropelike structure of great tensile strength. Indeed, the toughness of leather, chemically treated animal skin, is due to its content of collagen. Elastin is also a cross-linked protein, but, instead of forming rigid coils, it imparts elasticity to tissues. Only one type of elastin is known; it varies in elasticity according to variations in its cross-linking.

The adhesive proteins of the extracellular matrix bind matrix molecules to one another and to cell surfaces. These proteins are modular in that they contain several functional domains packaged together in a single molecule. Each domain binds to a specific matrix component or to a specific site on a cell. The major adhesive protein of the interstitial matrix is called fibronectin; that of the basal lamina is known as laminin.

Cell-matrix Interactions

Molecules intimately associated with the cell membrane link cells to the extracellular matrix. These molecules, called matrix receptors, bind selectively to specific matrix components and interact, directly or indirectly, with actin protein fibres that form the cytoskeleton inside the cell. This association of actin fibres with matrix components via receptors on the cell membrane can influence the organization of membrane molecules as well as matrix components and can modify the shape and function of the cytoskeleton. Changes in the cytoskeleton can lead to changes in cell shape, movement, metabolism, and development.

Intercellular Recognition and Cell Adhesion

The ability of cells to recognize and adhere to one another plays an important role in cell survival and reproduction. For example, when starved, several types of single-cell organisms band together to develop the specialized cells needed for reproduction. In this process, certain cells at the centre of the developing aggregate secrete chemicals that cause the other cells to adhere tightly into a group. In the case of slime mold amoebas, starvation causes the secretion of a compound, cyclic adenosine monophosphate (cyclic AMP, or CAMP), that induces the cells to stick together end to end. With further aggregation, the cells produce another cell-surface glycoprotein with which they stick to one another over their entire surfaces. The cellular aggregates then produce an extracellular matrix, which holds the cells together in a specific structural form.

Tissue and Species Recognition

Marine sponges are multicellular animals that can regenerate from single cells. The cells of a sponge rely on the processes of intercellular recognition and cellular adhesion to form aggregates of cells of the same species that eventually develop into an adult sponge.

Some multicellular animals or tissues can be dissociated into suspensions of single cells that show the same cellular recognition and adhesion as do aggregates of single-cell organisms. The marine sponge, for example, can be sieved through a mesh, yielding single cells and cells in clumps. When this cell suspension is rotated in culture, the cells reaggregate and in time reform a normal sponge. This reassociation shows selective cell recognition; that is, only cells of the same species reassociate. The ability of the cells to distinguish cells of their own species from those of others is mediated by proteoglycan molecules in the extracellular matrix. The proteoglycan binds to specific cell-surface receptor sites that are unique to a single species of sponge.

Cells from tissues of vertebrate animals can, like sponge cells, be dissociated and allowed to reaggregate. For example, when vertebrate embryonic cells from two different tissues are dissociated and then rotated together in culture, the cells form a multicellular aggregate within which they sort according to the type of tissue, a sorting that occurs regardless of whether the cells are from the same or different species. The specificity is due to a set of cell-surface glycoproteins called cell adhesion molecules (CAM). A portion of the CAM that extends from the surface of a cell adheres to identical molecules on the surface of adjacent cells. These CAM appear early in embryonic life, and their amounts in tissues change as the organs develop. The CAM, however, are not responsible for the stable adhesion of one cell to another; this more permanent adhesion is carried out by cell junctions.

Cell Junctions

There are three functional categories of cell junction: adhering junctions, often called desmosomes; tight, or occluding, junctions; and gap, or permeable, junctions. Adhering junctions hold cells together mechanically and are associated with intracellular fibres of the cytoskeleton. Tight junctions also hold cells together, but they form a nearly leakproof intercellular seal by fusion of adjacent cell membranes. Both adhering junctions and tight junctions are present primarily in epithelial cells. Many cell types also possess gap junctions, which allow small molecules to pass from one cell to the next through a channel.

Adhering Junctions

Cells subject to abrasion or other mechanical stress, such as those of the surface epithelia of the skin, have junctions that adhere cells to one another and to the extracellular matrix. These adhering junctions are called desmosomes when occurring between cells and hemidesmosomes (half-desmosomes) when linked to the matrix. Adhering junctions distribute mechanical shear force throughout the tissue and to the underlying matrix by virtue of their association with intermediate filaments crossing the interior of the cell. The linkage of these filaments, also called keratin filaments, to the desmosomes and, through these junctions, to adjacent cells provides a nearly continuous fibrous network throughout an epithelial sheet. Adhering junctions are also seen in other types of cells—for example, in the muscles of the heart and uterus—allowing these cells to remain anchored together despite the contractions of the muscles.

Tight Junctions

Sheets of cells separate fluids within the organs from fluids outside, as in the epithelial layer lining the intestine. This separation requires leakproof junctions between cells. Tight junctions form leakproof seals by fusing the plasma membranes of adjacent cells, creating a continuous barrier through which molecules cannot pass. The membranes are fused by tight associations of two types of specialized integral membrane proteins, in turn repelling large water-soluble molecules. In invertebrates this function is provided by septate junctions, in which the proteins of the membrane rather than the lipids form the seal.

Gap Junctions

These junctions allow communication between adjacent cells via the passage of small molecules directly from the cytoplasm of one cell to that of another. Molecules that can pass between cells coupled by gap junctions include inorganic salts, sugars, amino acids, nucleotides, and vitamins but not large molecules such as proteins or nucleic acids.

Gap junctions are crucial to the integration of certain cellular activities. For example, heart muscle cells generate electrical current by the movement of inorganic salts. If the cells are coupled, they will share this electrical current, allowing the synchronous contraction of all the cells in the tissue. This coupling function requires the regulation of molecular traffic through the gaps. The junctions are not open pores but dynamic channels, which change their permeability with changes in cellular activity. They consist of proteins completely crossing the cell membrane as six-sided columns with central pores. Under certain conditions the proteins are thought to change shape, causing the pores to become smaller or larger and thus changing the permeability of the junction.

Gap junctions are also found in tissues that are not electrically active. In these tissues, the junctions allow nutrients and waste products to travel throughout the tissue. Cells in such tissues are said to be metabolically coupled. During the formation of embryos, gap junctions are crucial to establishing differences between separate groups of cells, the coupled cells undergoing development together to become a specialized tissue.

Cell-to-Cell Communication via Chemical Signaling

In addition to cell-matrix and cell-cell interactions, cell behaviour in multicellular organisms is coordinated by the passage of chemical or electrical signals between cells. The most common form of chemical signaling is via molecules secreted from the cells and moving through the extracellular space. Signaling molecules may also remain on cell surfaces, influencing other cells only after the cells make physical contact. Finally, gap junctions allow small molecules to move between the cytoplasms of adjacent cells.

Types of Chemical Signaling

Chemical signals secreted by cells can act over varying distances. In the autocrine signaling process, molecules act on the same cells that produce them. In paracrine signaling, they act on nearby cells. Autocrine signals include extracellular matrix molecules and various factors that stimulate cell growth. An example of paracrine signals is the chemical transmitted from nerve to muscle that causes the muscle to contract. In this instance, the muscle cells have regions specialized to receive chemical signals from an adjacent nerve cell. In both autocrine and paracrine signaling, the chemical signal works in the immediate vicinity of the cell that produces it and is present at high concentrations. A chemical signal picked up by the bloodstream and taken to distant sites is called an endocrine signal. Most hormones produced in vertebrates are endocrine signals, such as the hormones produced in the pituitary gland at the base of the brain and carried by the bloodstream to act at low concentrations on the thyroid or adrenal glands.

The concentration at which a chemical signal acts has significance for its target cell. Chemical signals that act at high concentration act locally and rapidly. On the other hand, chemical signals that act at low concentrations act at distances and are generally slow.

Signal Receptors

The ability of a cell to respond to an extracellular signal depends on the presence of specific proteins called receptors, which are located on the cell surface or in the cytoplasm. Receptors bind chemical signals that ultimately trigger a mechanism to modify the behaviour of the target cell. Cells may contain an array of specific receptors that allow them to respond to a variety of chemical signals.

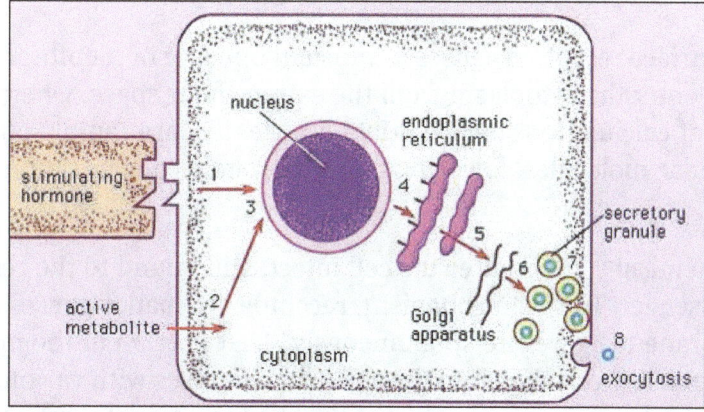

Hormones and active metabolites bind to different types of receptors. Water-soluble molecules (i.e., insulin) cannot pass through the lipid membrane of a cell and thus rely on cell surface receptors to transmit messages to the interior of the cell. In contrast, lipid-soluble molecules (i.e., certain active metabolites) are able to diffuse through the lipid membrane to communicate messages directly to the nucleus.

Signal molecules are either soluble or insoluble. Water-soluble molecules, such as the polypeptide hormone insulin, bind to receptors at cell surfaces. On the other hand, lipid-soluble molecules, such as the steroid hormones produced by the ovary or testis, pass through the lipid bilayer of the cell membrane to reach receptors within the cytoplasm. Extracellular matrix molecules are chemical signals, but, because of their size and insolubility, they act only on cell surface receptors and are neither taken up by the cells nor rapidly destroyed.

Cellular Response

The binding of chemical signals to their corresponding receptors induces events within the cell that ultimately change its behaviour. The nature of these intracellular events differs according to the type of receptor. Also, the same chemical signal can trigger different responses in different types of cell.

Cell surface receptors work in several ways when they are occupied. Some receptors enter the cell still bound to the chemical signal. Others activate membrane enzymes, which produce certain intracellular chemical mediators. Still other receptors open membrane channels, allowing a flow of ions that causes either a change in the electrical properties of the membrane or a change in the ion concentration in the cytoplasm. This regulation of enzymes or membrane channels produces changes in the concentration of intracellular signaling molecules, which are often called second messengers (the first messenger being the extracellular chemical signal bound to the receptor).

Two common intracellular signaling molecules are cyclic AMP and the calcium ion. Cyclic AMP is a derivative of adenosine triphosphate, the ubiquitous energy-carrying molecule of the cell. The intracellular concentrations of both cyclic AMP and calcium ions are normally very low. The binding of an extracellular chemical signal to a cell surface receptor stimulates an enzyme complex in the membrane to produce cyclic AMP. This second messenger then diffuses into the cytoplasm and acts on intracellular enzymes called kinases that modify the behaviour of the cell, culminating in the activation of target genes that increase the synthesis of certain proteins. The action of cyclic AMP is brief because it is rapidly degraded by specific enzymes.

Occupancy of other surface receptors causes a transient opening of membrane channels. This can allow calcium ions to enter the cytoplasm from the extracellular space, where their concentration is higher. The action of calcium ions is also brief because they are rapidly pumped out of the cell or bound to intracellular molecules, lowering the cytoplasmic concentration to the state existing before stimulation.

Some extracellular chemical signals enter the cell intact, still bound to the receptor, without generating a second messenger. In this mechanism, receptor occupancy causes individual receptors within the cell membrane to aggregate spontaneously. That portion of the membrane containing the aggregated receptors is then taken into the cell, where it fuses with various membrane-bounded organelles in the cytoplasm. In some instances the chemical signal is released within the organelles, and in almost all instances the ingested membrane is rapidly returned to the cell membrane along with the surface receptors.

Plant Cell Wall

The plant cell wall is a specialized form of extracellular matrix that surrounds every cell of a plant and is responsible for many of the characteristics distinguishing plant from animal cells. Although often perceived as an inactive product serving mainly mechanical and structural purposes, the cell wall actually has a multitude of functions upon which plant life depends. Such functions include: (1) providing the protoplast, or living cell, with mechanical protection and a chemically buffered environment, (2) providing a porous medium for the circulation and distribution of water, minerals, and other small nutrient molecules, (3) providing rigid building blocks from which stable structures of higher order, such as leaves and stems, can be produced, and (4) providing a storage site of regulatory molecules that sense the presence of pathogenic microbes and control the development of tissues.

Mechanical Properties of Wall Layers

All cell walls contain two layers, the middle lamella and the primary cell wall, and many cells produce an additional layer, called the secondary wall. The middle lamella serves as a cementing layer between the primary walls of adjacent cells. The primary wall is the cellulose-containing layer laid down by cells that are dividing and growing. To allow for cell wall expansion during growth, primary walls are thinner and less rigid than those of cells that have stopped growing. A fully grown plant cell may retain its primary cell wall (sometimes thickening it), or it may deposit an additional, rigidifying layer of different composition; this is the secondary wall. Secondary cell walls are responsible for most of the plant's mechanical support as well as the mechanical properties prized in wood. In contrast to the permanent stiffness and load-bearing capacity of thick secondary walls, the thin primary walls are capable of serving a structural, supportive role only when the vacuoles within the cell are filled with water to the point that they exert a turgor pressure against the cell wall. Turgor-induced stiffening of primary walls is analogous to the stiffening of the sides of a pneumatic tire by air pressure. The wilting of flowers and leaves is caused by a loss of turgor pressure, which results in turn from the loss of water from the plant cells.

Components of the Cell Wall

Although primary and secondary wall layers differ in detailed chemical composition and structural organization, their basic architecture is the same, consisting of cellulose fibres of great tensile strength embedded in a water-saturated matrix of polysaccharides and structural glycoproteins.

Cellulose

Cellulose consists of several thousand glucose molecules linked end to end. The chemical links between the individual glucose subunits give each cellulose molecule a flat ribbonlike structure that allows adjacent molecules to band laterally together into microfibrils with lengths ranging from two to seven micrometres. Cellulose fibrils are synthesized by enzymes floating in the cell membrane and are arranged in a rosette configuration. Each rosette appears capable of "spinning" a microfibril into the cell wall. During this process, as new glucose subunits are added to the growing end of the fibril, the rosette is pushed around the cell on the surface of the cell membrane, and its cellulose fibril becomes wrapped around the protoplast. Thus, each plant cell can be viewed as making its own cellulose fibril cocoon.

Matrix Polysaccharides

The two major classes of cell wall matrix polysaccharides are the hemicelluloses and the pectic polysaccharides, or pectins. Both are synthesized in the Golgi apparatus, brought to the cell surface in small vesicles, and secreted into the cell wall.

Hemicelluloses consist of glucose molecules arranged end to end as in cellulose, with short side chains of xylose and other uncharged sugars attached to one side of the ribbon. The other side of the ribbon binds tightly to the surface of cellulose fibrils, thereby coating the microfibrils with hemicellulose and preventing them from adhering together in an uncontrolled manner. Hemicellulose molecules have been shown to regulate the rate at which primary cell walls expand during growth.

The heterogeneous, branched, and highly hydrated pectic polysaccharides differ from hemicelluloses in important respects. Most notably, they are negatively charged because of galacturonic acid residues, which, together with rhamnose sugar molecules, form the linear backbone of all pectic polysaccharides. The backbone contains stretches of pure galacturonic acid residues interrupted by segments in which galacturonic acid and rhamnose residues alternate; attached to these latter segments are complex, branched sugar side chains. Because of their negative charge, pectic polysaccharides bind tightly to positively charged ions, or cations. In cell walls, calcium ions cross-link the stretches of pure galacturonic acid residues tightly, while leaving the rhamnose-containing segments in a more open, porous configuration. This cross-linking creates the semirigid gel properties characteristic of the cell wall matrix—a process exploited in the preparation of jellied preserves.

Proteins

Although plant cell walls contain only small amounts of protein, they serve a number of important functions. The most prominent group are the hydroxyproline-rich glycoproteins, shaped like rods with connector sites, of which extensin is a prominent example. Extensin contains 45 percent hydroxyproline and 14 percent serine residues distributed along its length. Every hydroxyproline residue carries a short side chain of arabinose sugars, and most serine residues carry a galactose sugar; this gives rise to long molecules, resembling bottle brushes, that are secreted into the cell wall toward the end of primary-wall formation and become covalently cross-linked into a mesh at the time that cell growth stops. Plant cells may control their ultimate size by regulating the time at which this cross-linking of extensin molecules occurs.

In addition to the structural proteins, cell walls contain a variety of enzymes. Most notable are those that cross-link extensin, lignin, cutin, and suberin molecules into networks. Other enzymes help protect plants against fungal pathogens by breaking fragments off of the cell walls of the fungi. The fragments in turn induce defense responses in underlying cells. The softening of ripe fruit and dropping of leaves in the autumn are brought about by cell wall-degrading enzymes.

Plastics

Cell wall plastics such as lignin, cutin, and suberin all contain a variety of organic compounds cross-linked into tight three-dimensional networks that strengthen cell walls and make them more resistant to fungal and bacterial attack. Lignin is the general name for a diverse group of polymers of aromatic alcohols. Deposited mostly in secondary cell walls and providing the rigidity of terrestrial vascular plants, it accounts for up to 30 percent of a plant's dry weight. The diversity of cross-links between the polymers—and the resulting tightness—makes lignin a formidable barrier to the penetration of most microbes.

Agave shawii (top) and Echeveria (bottom), two types of xerophytes (plants adapted to arid habitats). They develop highly cutinized fleshy leaves and stems for water storage with which they modulate the effects of strong sunlight, low humidity, and scant water.

Cutin and suberin are complex biopolyesters composed of fatty acids and aromatic compounds. Cutin is the major component of the cuticle, the waxy, water-repelling surface layer of cell walls exposed to the environment aboveground. By reducing the wetability of leaves and stems—and thereby affecting the ability of fungal spores to germinate—it plays an important part in the defense strategy of plants. Suberin serves with waxes as a surface barrier of underground parts. Its synthesis is also stimulated in cells close to wounds, thereby sealing off the wound surfaces and protecting underlying cells from dehydration.

Intercellular Communication

Plasmodesmata

Similar to the gap junction of animal cells is the plasmodesma, a channel passing through the cell wall and allowing direct molecular communication between adjacent plant cells. Plasmodesmata are lined with cell membrane, in effect uniting all connected cells with one continuous cell membrane. Running down the middle of each channel is a thin membranous tube that connects the endoplasmic reticula (ER) of the two cells. This structure is a remnant of the ER of the original parent cell, which, as the parent cell divided, was caught in the developing cell plate.

Although the precise mechanisms are not fully understood, the plasmodesma is thought to regulate the passage of small molecules such as salts, sugars, and amino acids by constricting or dilating the openings at each end of the channel.

Oligosaccharides with Regulatory Functions

The discovery of cell wall fragments with regulatory functions opened a new era in plant research. For years scientists had been puzzled by the chemical complexity of cell wall polysaccharides, which far exceeds the structural requirements of plant cell walls. The answer came when it was found that specific fragments of cell wall polysaccharides, called oligosaccharins, are able to induce specific responses in plant cells and tissues. One such fragment, released by enzymes used by fungi to break down plant cell walls, consists of a linear polymer of 10 to 12 galacturonic acid residues. Exposure of plant cells to such fragments induces them to produce antibiotics known as phytoalexins. In other experiments it has been shown that exposing strips of tobacco stem cells to a different type of cell wall fragment leads to the growth of roots; other fragments lead to the formation of stems, and yet others to the production of flowers. In all instances the concentration of oligosaccharins required to bring about the observed responses is equal to that of hormones in animal cells; indeed, oligosaccharins may be viewed as the oligosaccharide hormones of plants.

Cell Division and Growth

In unicellular organisms, cell division is the means of reproduction; in multicellular organisms, it is the means of tissue growth and maintenance. Survival of the eukaryotes depends upon interactions between many cell types, and it is essential that a balanced distribution of types be maintained. This is achieved by the highly regulated process of cell proliferation. The growth and division of different cell populations are regulated in different ways, but the basic mechanisms are similar throughout multicellular organisms.

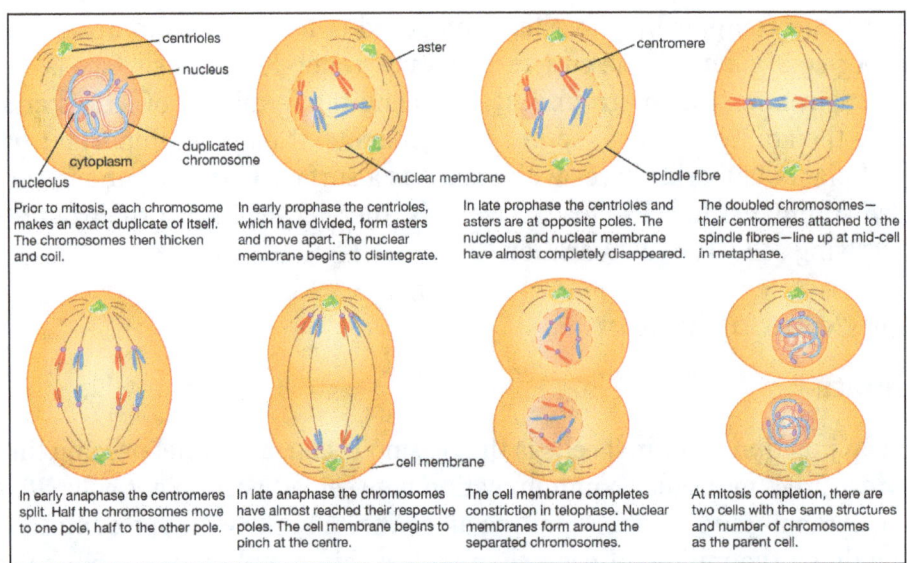

One cell gives rise to two genetically identical daughter cells during the process of mitosis.

Most tissues of the body grow by increasing their cell number, but this growth is highly regulated to maintain a balance between different tissues. In adults most cell division is involved in tissue renewal rather than growth, many types of cells undergoing continuous replacement. Skin cells, for example, are constantly being sloughed off and replaced; in this case, the mature differentiated cells do not divide, but their population is renewed by division of immature stem cells. In certain other cells, such as those of the liver, mature cells remain capable of division to allow growth or regeneration after injury.

In contrast to these patterns, other types of cells either cannot divide or are prevented from dividing by certain molecules produced by nearby cells. As a result, in the adult organism, some tissues have a greatly reduced capacity to renew damaged or diseased cells. Examples of such tissues include heart muscle, nerve cells of the central nervous system, and lens cells in mammals. Maintenance and repair of these cells is limited to replacing intracellular components rather than replacing entire cells.

Duplication of the Genetic Material

Before a cell can divide, it must accurately and completely duplicate the genetic information encoded in its DNA in order for its progeny cells to function and survive. This is a complex problem because of the great length of DNA molecules. Each human chromosome consists of a long double spiral, or helix, each strand of which consists of more than 100 million nucleotides.

The duplication of DNA is called DNA replication, and it is initiated by complex enzymes called DNA polymerases. These progress along the molecule, reading the sequences of nucleotides that are linked together to make DNA chains. Each strand of the DNA double helix, therefore, acts as a template specifying the nucleotide structure of a new growing chain. After replication, each of the two daughter DNA double helices consists of one parental DNA strand wound around one newly synthesized DNA strand.

In order for DNA to replicate, the two strands must be unwound from each other. Enzymes called helicases unwind the two DNA strands, and additional proteins bind to the separated strands to stabilize them and prevent them from pairing again. In addition, a remarkable class of enzyme called DNA topoisomerase removes the helical twists by cutting either one or both strands and then resealing the cut. These enzymes can also untangle and unknot DNA when it is tightly coiled into a chromatin fibre.

In the circular DNA of prokaryotes, replication starts at a unique site called the origin of replication and then proceeds in both directions around the molecule until the two processes meet, producing two daughter molecules. In rapidly growing prokaryotes, a second round of replication can start before the first has finished. The situation in eukaryotes is more complicated, as replication moves more slowly than in prokaryotes. At 500 to 5,000 nucleotides per minute (versus 100,000 nucleotides per minute in prokaryotes), it would take a human chromosome about a month to replicate if started at a single site. Actually, replication begins at many sites on the long chromosomes of animals, plants, and fungi. Distances between adjacent initiation sites are not always the same; for example, they are closer in the rapidly dividing embryonic cells of frogs or flies than in adult cells of the same species.

Accurate DNA replication is crucial to ensure that daughter cells have exact copies of the genetic information for synthesizing proteins. Accuracy is achieved by a "proofreading" ability of the DNA polymerase itself. It can erase its own errors and then synthesize anew. There are also repair systems that correct genetic damage to DNA. For example, the incorporation of an incorrect nucleotide, or damage caused by mutagenic agents, can be corrected by cutting out a section of the daughter strand and recopying the parental strand.

Cell Division

Mitosis and Cytokinesis

In eukaryotes the processes of DNA replication and cell division occur at different times of the cell division cycle. During cell division, DNA condenses to form short, tightly coiled, rodlike chromosomes. Each chromosome then splits longitudinally, forming two identical chromatids. Each pair of chromatids is divided between the two daughter cells during mitosis, or division of the nucleus, a process in which the chromosomes are propelled by attachment to a bundle of microtubules called the mitotic spindle.

Mitosis can be divided into five phases. In prophase the mitotic spindle forms and the chromosomes condense. In prometaphase the nuclear envelope breaks down (in many but not all eukaryotes) and the chromosomes attach to the mitotic spindle. Both chromatids of each chromosome attach to the spindle at a specialized chromosomal region called the kinetochore. In metaphase the condensed chromosomes align in a plane across the equator of the mitotic spindle. Anaphase follows as the separated chromatids move abruptly toward opposite spindle poles. Finally, in telophase a new nuclear envelope forms around each set of unraveling chromatids.

An essential feature of mitosis is the attachment of the chromatids to opposite poles of the mitotic spindle. This ensures that each of the daughter cells will receive a complete set of chromosomes. The mitotic spindle is composed of microtubules, each of which is a tubular assembly of molecules of the protein tubulin. Some microtubules extend from one spindle pole to the other, while a second class extends from one spindle pole to a chromatid. Microtubules can grow or shrink by the addition or removal of tubulin molecules. The shortening of spindle microtubules at anaphase propels attached chromatids to the spindle poles, where they unravel to form new nuclei.

The two poles of the mitotic spindle are occupied by centrosomes, which organize the microtubule arrays. In animal cells each centrosome contains a pair of cylindrical centrioles, which are themselves composed of complex arrays of microtubules. Centrioles duplicate at a precise time in the cell division cycle, usually close to the start of DNA replication.

After mitosis comes cytokinesis, the division of the cytoplasm. This is another process in which animal and plant cells differ. In animal cells cytokinesis is achieved through the constriction of the cell by a ring of contractile microfilaments consisting of actin and myosin, the proteins involved in muscle contraction and other forms of cell movement. In plant cells the cytoplasm is divided by the formation of a new cell wall, called the cell plate, between the two daughter cells. The cell plate arises from small Golgi-derived vesicles that coalesce in a plane across the equator of the late telophase spindle to form a disk-shaped structure. In this process, each vesicle contributes its membrane to the forming cell membranes and its matrix contents to the forming cell wall. A second set of vesicles extends the edge of the cell plate until it reaches and fuses with the sides of the parent cell, thereby completely separating the two new daughter cells. At this point, cellulose synthesis commences, and the cell plate becomes a primary cell wall.

Meiosis

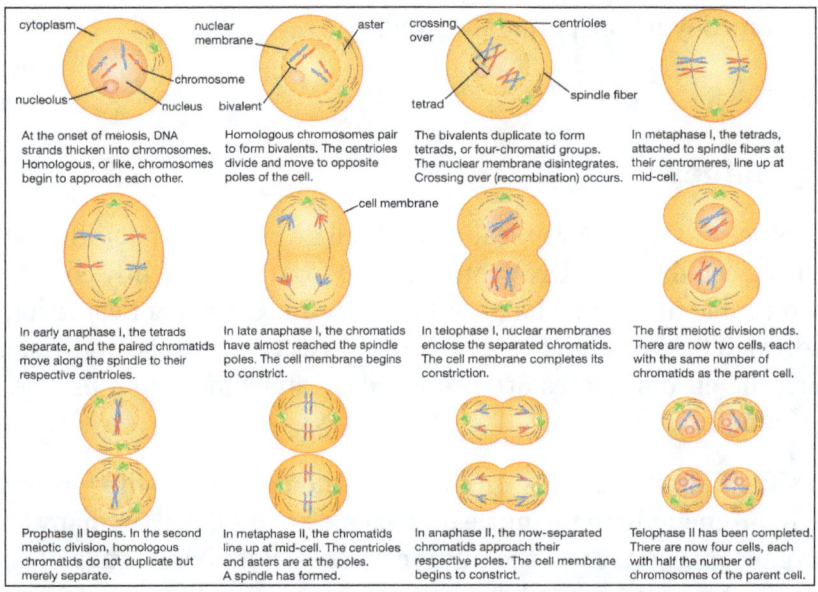

The formation of gametes (sex cells) occurs during the process of meiosis.

A specialized division of chromosomes called meiosis occurs during the formation of the reproductive cells, or gametes, of sexually reproducing organisms. Gametes such as ova, sperm, and pollen begin as germ cells, which, like other types of cells, have two copies of each gene in their nuclei. The chromosomes composed of these matching genes are called homologs. During DNA replication, each chromosome duplicates into two attached chromatids. The homologous chromosomes are then separated to opposite poles of the meiotic spindle by microtubules similar to those of the mitotic spindle. At this stage in the meiosis of germ cells, there is a crucial difference from the mitosis of other cells. In meiosis the two chromatids making up each chromosome remain together, so that whole chromosomes are separated from their homologous partners. Cell division then occurs, followed by a second division that resembles mitosis more closely in that it separates the two chromatids of each remaining chromosome. In this way, when meiosis is complete, each mature gamete receives only one copy of each gene instead of the two copies present in other cells.

Cell Division Cycle

In prokaryotes, DNA synthesis can take place uninterrupted between cell divisions, and new cycles of DNA synthesis can begin before previous cycles have finished. In contrast, eukaryotes duplicate their DNA exactly once during a discrete period between cell divisions. This period is called the S (for synthetic) phase. It is preceded by a period called G1 (meaning "first gap") and followed by a period called G_2, during which nuclear DNA synthesis does not occur.

The four periods G_1, S, G_2, and M (for mitosis) make up the cell division cycle. The cell cycle characteristically lasts between 10 and 20 hours in rapidly proliferating adult cells, but it can be arrested for weeks or months in quiescent cells or for a lifetime in neurons of the brain. Prolonged arrest of this type usually occurs during the G_1 phase and is sometimes referred to as G_0. In contrast, some embryonic cells, such as those of fruit flies (vinegar flies), can complete entire cycles and divide in only 11 minutes. In these exceptional cases, G_1 and G_2 are undetectable, and mitosis alternates with DNA synthesis. In addition, the duration of the S phase varies dramatically. The fruit fly embryo takes only four minutes to replicate its DNA, compared with several hours in adult cells of the same species.

Controlled Proliferation

Several studies have identified the transition from the G_1 to the S phase as a crucial control point of the cell cycle. Stimuli are known to cause resting cells to proliferate by inducing them to leave G1 and begin DNA synthesis. These stimuli, called growth factors, are naturally occurring proteins specific to certain groups of cells in the body. They include nerve growth factor, epidermal growth factor, and platelet-derived growth factor. Such factors may have important roles in the healing of wounds as well as in the maintenance and growth of normal tissues. Many growth factors are known to act on the external membrane of the cell, by interacting with specialized protein receptor molecules. These respond by triggering further cellular changes, including an increase in calcium levels that makes the cell interior more alkaline and the addition of phosphate groups to the amino acid tyrosine in proteins. The complex response of cells to growth factors is of fundamental importance to the control of cell proliferation.

Failure of Proliferation Control

Cancer can arise when the controlling factors over cell growth fail and allow a cell and its descendants to keep dividing at the expense of the organism. Studies of viruses that transform cultured cells and thus lead to the loss of control of cell growth have provided insight into the mechanisms that drive the formation of tumours. Transformed cells may differ from their normal progenitors by continuing to proliferate at very high densities, in the absence of growth factors, or in the absence of a solid substrate for support.

Retroviral insertion can convert a proto-oncogene, integral to the control of cell division, into an oncogene, the agent responsible for transforming a healthy cell into a cancer cell. An acutely transforming retrovirus (shown at top in above figure), which produces tumours within weeks of infection, incorporates genetic material from a host cell into its own genome upon infection, forming a viral oncogene. When the viral oncogene infects another cell, an enzyme called reverse transcriptase copies the single-stranded genetic material into double-stranded DNA, which is then integrated into the cellular genome. A slowly transforming retrovirus (shown at bottom in above figure), which requires months to elicit tumour growth, does not disrupt cellular function through the insertion of a viral oncogene. Rather, it carries a promoter gene that is integrated into the cellular genome of the host cell next to or within a proto-oncogene, allowing conversion of the proto-oncogene to an oncogene.

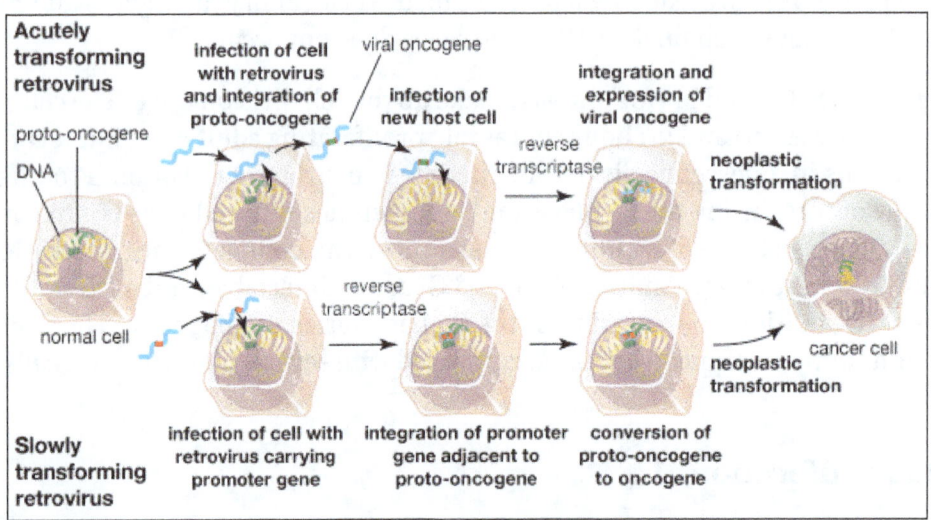

Cancer-causing retroviruses.

Major advances in the understanding of growth control have come from studies of the viral genes that cause transformation. These viral oncogenes have led to the identification of related cellular genes called protooncogenes. Protooncogenes can be altered by mutation or epigenetic modification, which converts them into oncogenes and leads to cell transformation. Specific oncogenes are activated in particular human cancers. For example, an oncogene called RAS is associated with many epithelial cancers, while another, called MYC, is associated with leukemias.

An interesting feature of oncogenes is that they may act at different levels corresponding to the multiple steps seen in the development of cancer. Some oncogenes immortalize cells so that they divide indefinitely, whereas normal cells die after a limited number of generations. Other oncogenes transform cells so that they grow in the absence of growth factors. A combination of these two functions leads to loss of proliferation control, whereas each of these functions on its own cannot. The mode of action of oncogenes also provides important clues to the nature of growth control and cancer. For example, some oncogenes are known to encode receptors for growth factors that may cause continuous proliferation in the absence of appropriate growth factors.

Loss of growth control has the added consequence that cells no longer repair their DNA effectively, and thus aberrant mitoses occur. As a result, additional mutations arise that subvert a cell's normal constraints to remain in its tissue of origin. Epithelial tumour cells, for example, acquire the ability to cross the basal lamina and enter the bloodstream or lymphatic system, where they migrate to other parts of the body, a process called metastasis. When cells metastasize to distant tissues, the tumour is described as malignant, whereas prior to metastasis a tumour is described as benign.

Cell Differentiation

Adult organisms are composed of a number of distinct cell types. Cells are organized into tissues, each of which typically contains a small number of cell types and is devoted to a specific physiological function. For example, the epithelial tissue lining the small intestine contains columnar absorptive cells, mucus-secreting goblet cells, hormone-secreting endocrine cells, and enzyme-secreting Paneth cells. In addition, there exist undifferentiated dividing cells that lie in the crypts between the intestinal villi and serve to replace the other cell types when they become damaged or worn out. Another example of a differentiated tissue is the skeletal tissue of a long bone, which contains osteoblasts (large cells that synthesize bone) in the outer sheath and osteocytes (mature bone cells) and osteoclasts (multinucleate cells involved in bone remodeling) within the matrix.

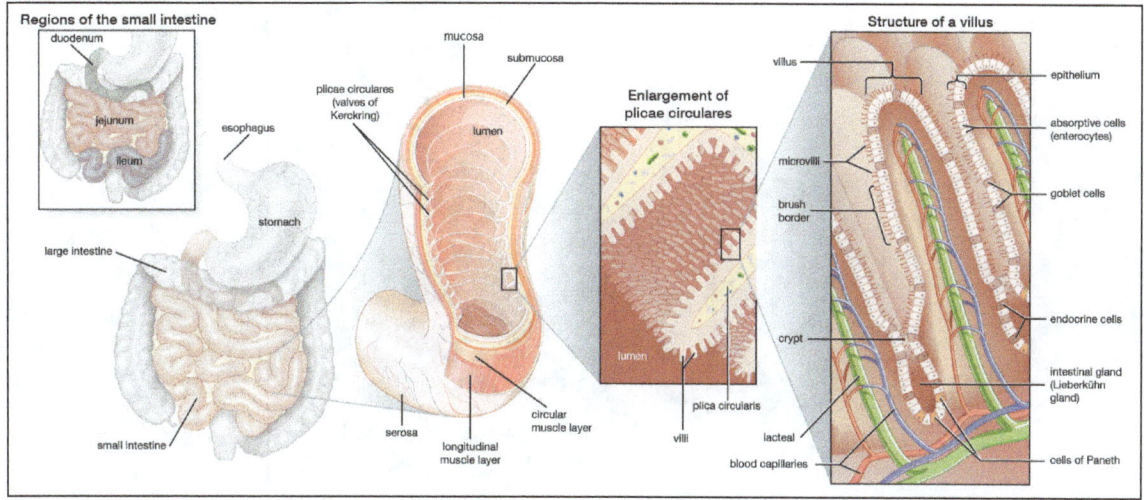

The small intestine contains many distinct types of cells, each of which serves a specific function.

In general, the simpler the overall organization of the animal, the fewer the number of distinct cell types that they possess. Mammals contain more than 200 different cell types, whereas simple invertebrate animals may have only a few different types. Plants are also made up of differentiated cells, but they are quite different from the cells of animals. For example, a leaf in a higher plant is covered with a cuticle layer of epidermal cells. Among these are pores composed of two specialized cells, which regulate gaseous exchange across the epidermis. Within the leaf is the mesophyll, a spongy tissue responsible for photosynthetic activity. There are also veins composed of xylem elements, which transport water up from the soil, and phloem elements, which transport products of photosynthesis to the storage organs.

The epidermis is often covered with a waxy protective cuticle that helps prevent water loss from inside the leaf. Oxygen, carbon dioxide, and water enter and exit the leaf through pores (stomata) scattered mostly along the lower epidermis. The stomata are opened and closed by the contraction and expansion of surrounding guard cells. The vascular, or conducting, tissues are known as xylem and phloem; water and minerals travel up to the leaves from the roots through the xylem, and sugars made by photosynthesis are transported to other parts of the plant through the phloem. Photosynthesis occurs within the chloroplast-containing mesophyll layer.

The various cell types have traditionally been recognized and classified according to their appearance in the light microscope following the process of fixing, processing, sectioning, and staining tissues that is known as histology. Classical histology has been augmented by a variety of more discriminating techniques. Electron microscopy allows for higher magnifications. Histochemistry involves the use of coloured precipitating substrates to stain particular enzymes in situ. Immuno-histochemistry uses specific antibodies to identify particular substances, usually proteins or carbohydrates, within cells. In situ hybridization involves the use of nucleic acid probes to visualize the location of specific messenger RNAs (mRNA). These modern methods have allowed the identification of more cell types than could be visualized by classical histology, particularly in the brain, the immune system, and among the hormone-secreting cells of the endocrine system.

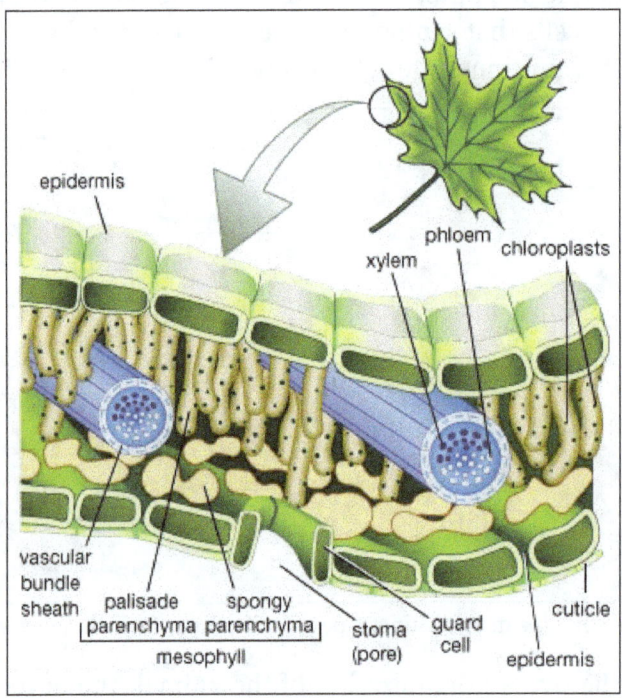

Structures of a leaf.

Differentiated State

The biochemical basis of cell differentiation is the synthesis by the cell of a particular set of proteins, carbohydrates, and lipids. This synthesis is catalyzed by proteins called enzymes. Each enzyme in turn is synthesized in accordance with a particular gene, or sequence of nucleotides in the DNA of the cell nucleus. A particular state of differentiation, then, corresponds to the set of genes that is expressed and the level to which it is expressed.

It is believed that all of an organism's genes are present in each cell nucleus, no matter what the cell type, and that differences between tissues are not due to the presence or absence of certain genes but are due to the expression of some and the repression of others. In animals the best evidence for retention of the entire set of genes comes from whole animal cloning experiments in which the nucleus of a differentiated cell is substituted for the nucleus of a fertilized egg. In many species this can result in the development of a normal embryo that contains the full range of body parts and cell types. Likewise, in plants it is often possible to grow complete embryos from individual cells in tissue culture. Such experiments show that any nucleus has the genetic information required for the growth of a developing organism, and they strongly suggest that, for most tissues, cell differentiation arises from the regulation of genetic activity rather than the removal or destruction of unwanted genes. The only known exception to this rule comes from the immune system, where segments of DNA in developing white b

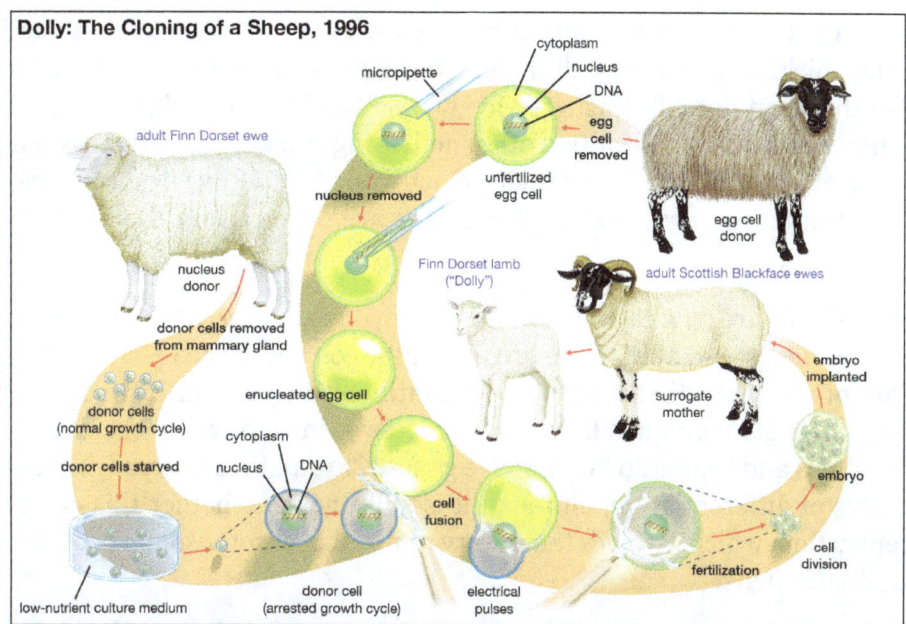

Dolly the sheep was successfully cloned in 1996 by fusing the nucleus from a mammary-gland cell of a Finn Dorset ewe into an enucleated egg cell taken from a Scottish Blackface ewe. Carried to term in the womb of another Scottish Blackface ewe, Dolly was a genetic copy of the Finn Dorset ewe.

lood cells are slightly rearranged, producing a wide variety of antibody and receptor molecules.

At the molecular level there are many ways in which the expression of a gene can be differentially regulated in different cell types. There may be differences in the copying, or transcription, of the gene into RNA; in the processing of the initial RNA transcript into mRNA; in the control of mRNA movement to the cytoplasm; in the translation of mRNA to protein; or in the stability of mRNA. However, the control of transcription has the most influence over gene expression and has received the most detailed analysis.

The DNA in the cell nucleus exists in the form of chromatin, which is made up of DNA bound to histones (simple alkaline proteins) and other nonhistone proteins. Most of the DNA is complexed into repeating structures called nucleosomes, each of which contains eight molecules of histone. Active genes are found in parts of the DNA where the chromatin has an "open" configuration, in which regulatory proteins are able to gain access to the DNA. The degree to which the chromatin opens depends on chemical modifications of the outer parts of the histone molecules and on the presence or absence of particular nonhistone proteins. Transcriptional control is exerted with the help of regulatory sequences that are found associated with a gene, such as the promoter sequence, a region near the start of the gene, and enhancer sequences, regions that lie elsewhere within the DNA that augment the activity of enzymes involved in the process of transcription. Whether or not transcription occurs depends on the binding of transcription factors to these regulatory sequences. Transcription factors are proteins that usually possess a DNA-binding region, which recognizes the specific regulatory sequence in the DNA, and an effector region, which activates or inhibits transcription. Transcription factors often work by recruiting enzymes that add modifications (e.g., acetyl groups or methyl groups) to or remove modifications from the outer parts of the histone molecules. This controls the folding of the chromatin and the accessibility of the DNA to RNA polymerase and other transcription factors.

In general, it requires several transcription factors working in combination to activate a gene. For example, the chicken delta 1 crystallin gene, normally expressed only in the lens of the eye, has a promoter that contains binding sites for two activating transcription factors and an enhancer that contains binding sites for two other activating transcription factors. There is also an additional enhancer site that can bind either an activator (deltaEF3) or a repressor (deltaEF1). Successful transcription requires that all these sites are occupied by the correct transcription factors.

Fully differentiated cells are qualitatively different from one another. States of terminal differentiation are stable and persistent, both in the lifetime of the cell and in successive cell generations (in the case of differentiated types that are capable of continued cell division). The inherent stability of the differentiated state is maintained by various processes, including feedback activation of genes by their own products and repression of inactive genes. Chromatin structure may be important in maintaining states of differentiation, although it is still unclear whether this can be maintained during DNA replication, which involves temporary removal of chromosomal proteins and unwinding of the DNA double helix.

A type of differentiation control that is maintained during DNA replication is the methylation of DNA, which tends to recruit histone deacetylases and hence close up the structure of the chromatin. DNA methylation occurs when a methyl group is attached to the exterior, or sugar-phosphate side, of a cytosine (C) residue. Cytosine methylation occurs only on a C nucleotide when it is connected to a G (guanine) nucleotide on the same strand of DNA. These nucleotide pairings are called CG dinucleotides. One class of DNA methylase enzyme can introduce new methylations when required, whereas another class, called maintenance methylases, methylates CG dinucleotides in the DNA double helix only when the CG of the complementary strand is already methylated. Each time the methylated DNA is replicated, the old strand has the methyl groups and the new strand does not. The maintenance methylase will then add methyl groups to all the CGs opposite the existing methyl groups to restore a fully methylated double helix. This mechanism guarantees stability of the DNA methylation pattern, and hence the differentiated state, during the processes of DNA replication and cell division.

Process of Differentiation

Differentiation from visibly undifferentiated precursor cells occurs during embryonic development, during metamorphosis of larval forms, and following the separation of parts in asexual reproduction. It also takes place in adult organisms during the renewal of tissues and the regeneration of missing parts. Thus, cell differentiation is an essential and ongoing process at all stages of life.

The visible differentiation of cells is only the last of a progressive sequence of states. In each state, the cell becomes increasingly committed toward one type of cell into which it can develop. States of commitment are sometimes described as "specification" to represent a reversible type of commitment and as "determination" to represent an irreversible commitment. Although states of specification and determination both represent differential gene activity, the properties of embryonic cells are not necessarily the same as those of fully differentiated cells. In particular, cells in specification states are usually not stable over prolonged periods of time.

Two mechanisms bring about altered commitments in the different regions of the early embryo: cytoplasmic localization and induction. Cytoplasmic localization is evident in the earliest stages of development of the embryo. During this time, the embryo divides without growth, undergoing cleavage divisions that produce separate cells called blastomeres. Each blastomere inherits a certain region of the original egg cytoplasm, which may contain one or more regulatory substances called cytoplasmic determinants. When the embryo has become a solid mass of blastomeres (called a morula), it generally consists of two or more differently committed cell populations—a result of the blastomeres having incorporated different cytoplasmic determinants. Cytoplasmic determinants may consist of mRNA or protein in a particular state of activation. An example of the influence of a cytoplasmic determinant is a receptor called Toll, located in the membranes of Drosophila (fruit fly) eggs. Activation of Toll ensures that the blastomeres will develop into ventral (underside) structures, while blastomeres containing inactive Toll will produce cells that will develop into dorsal (back) structures.

In induction, the second mechanism of commitment, a substance secreted by one group of cells alters the development of another group. In early development, induction is usually instructive; that is, the tissue assumes a different state of commitment in the presence of the signal than it would in the absence of the signal. Inductive signals often take the form of concentration gradients of substances that evoke a number of different responses at different concentrations. This leads to the formation of a sequence of groups of cells, each in a different state of specification. For example, in Xenopus (clawed frog) the early embryo contains a signaling centre called the organizer that secretes inhibitors of bone morphogenetic proteins (BMPs), leading to a ventral-to-dorsal (belly-to-back) gradient of BMP activity. The activity of BMP in the ventral region of the embryo suppresses the expression of transcription factors involved in the formation of the central nervous system and segmented muscles. Suppression ensures that these structures are formed only on the dorsal side, where there is decreased activity of BMP.

The final stage of differentiation often involves the formation of several types of differentiated cells from one precursor or stem cell population. Terminal differentiation occurs not only in embryonic development but also in many tissues in postnatal life. Control of this process depends on a system of lateral inhibition in which cells that are differentiating along a particular pathway send out signals that repress similar differentiation by their neighbours. For example, in the developing central nervous system of vertebrates, neurons arise from a simple tube of neuroepithelium, the cells of which possess a surface receptor called Notch. These cells also possess another cell surface molecule called Delta that can bind to and activate Notch on adjacent cells. Activation of Notch initiates a cascade of intracellular events that results in suppression of Delta production and suppression of neuronal differentiation. This means that the neuroepithelium generates only a few cells with high expression of Delta surrounded by a larger number of cells with low expression of Delta. High Delta production and low Notch activation makes the cells develop into neurons. Low Delta production and high Notch activation makes the cells remain as precursor cells or become glial (supporting) cells. A similar mechanism is known to produce the endocrine cells of the pancreas and the goblet cells of the intestinal epithelium. Such lateral inhibition systems work because cells in a population are never quite identical to begin with. There are always small differences, such as in the number of Delta molecules displayed on the cell surface. The mechanism of lateral inhibition amplifies these small differences, using them to bring about differential gene expression that leads to stable and persistent states of cell differentiation.

Errors in Differentiation

Three classes of abnormal cell differentiation are dysplasia, metaplasia, and anaplasia. Dysplasia indicates an abnormal arrangement of cells, usually arising from a disturbance in their normal growth behaviour. Some dysplasias are precursor lesions to cancer, whereas others are harmless and regress spontaneously. For example, dysplasia of the uterine cervix, called cervical intraepithelial neoplasia (CIN), may progress to cervical cancer. It can be detected by cervical smear cytology tests (Pap smears).

Metaplasia is the conversion of one cell type into another. In fact, it is not usually the differentiated cells themselves that change but rather the stem cell population from which they are derived. Metaplasia commonly occurs where chronic tissue damage is followed by extensive regeneration. For example, squamous metaplasia of the bronchi occurs when the ciliated respiratory epithelial cells of people who smoke develop into squamous, or flattened, cells. In intestinal metaplasia of the stomach, patches resembling intestinal tissue arise in the gastric mucosa, often in association with gastric ulcers. Both of these types of metaplasia may progress to cancer.

Anaplasia is a loss of visible differentiation that can occur in advanced cancer. In general, early cancers resemble their tissue of origin and are described and classified by their pattern of differentiation. However, as they develop, they produce variants of more abnormal appearance and increased malignancy. Finally, a highly anaplastic growth can occur, in which the cancerous cells bear no visible relation to the parent tissue.

Evolution of Cells

Development of Genetic Information

Life on Earth could not exist until a collection of catalysts appeared that could promote the synthesis of more catalysts of the same kind. Early stages in the evolutionary pathway of cells presumably centred on RNA molecules, which not only present specific catalytic surfaces but also contain the potential for their own duplication through the formation of a complementary RNA molecule. It is assumed that a small RNA molecule eventually appeared that was able to catalyze its own duplication.

Imperfections in primitive RNA replication likely gave rise to many variant autocatalytic RNA molecules. Molecules of RNA that acquired variations that increased the speed or the fidelity of self-replication would have outmultiplied other, less-competent RNA molecules. In addition, other small RNA molecules that existed in symbiosis with autocatalytic RNA molecules underwent natural selection for their ability to catalyze useful secondary reactions such as the production of better precursor molecules. In this way, sophisticated families of RNA catalysts could have evolved together, since cooperation between different molecules produced a system that was much more effective at self-replication than a collection of individual RNA catalysts.

Another major step in the evolution of the cell would have been the development, in one family of self-replicating RNA, of a primitive mechanism of protein synthesis. Protein molecules cannot provide the information for the synthesis of other protein molecules like themselves. This information must ultimately be derived from a nucleic acid sequence. Protein synthesis is much more complex than RNA synthesis, and it could not have arisen before a group of powerful RNA catalysts evolved. Each of these catalysts presumably has its counterpart among the RNA molecules that function in the current cell: (1) there was an information RNA molecule, much like messenger RNA (mRNA), whose nucleotide sequence was read to create an amino acid sequence; (2) there was a group of adaptor RNA molecules, much like transfer RNA (tRNA), that could bind to both mRNA and a specific activated amino acid; and (3) finally, there was an RNA catalyst, much like ribosomal RNA (rRNA), that facilitated the joining together of the amino acids aligned on the mRNA by the adaptor RNA.

At some point in the evolution of biological catalysts, the first cell was formed. This would have required the partitioning of the primitive biological catalysts into individual units, each surrounded by a membrane. Membrane formation might have occurred quite simply, since many amphiphilic molecules—half hydrophobic (water-repelling) and half hydrophilic (water-loving)—aggregate to form bilayer sheets in which the hydrophobic portions of the molecules line up in rows to form the interior of the sheet and leave the hydrophilic portions to face the water. Such bilayer sheets can spontaneously close up to form the walls of small, spherical vesicles, as can the phospholipid bilayer membranes of present-day cells.

As soon as the biological catalysts became compartmentalized into small individual units, or cells, the units would have begun to compete with one another for the same resources. The active competition that ensued must have greatly accelerated evolutionary change, serving as a powerful force for the development of more efficient cells. In this way, cells eventually arose that contained new catalysts, enabling them to use simpler, more abundant precursor molecules for their growth. Because these cells were no longer dependent on preformed ingredients for their survival, they were able to spread far beyond the limited environments where the first primitive cells arose.

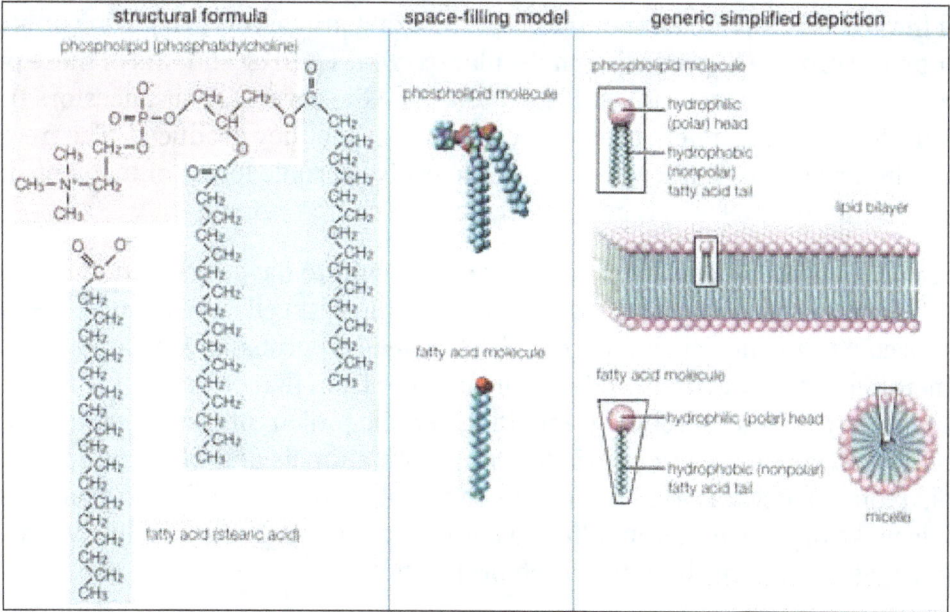

Structure and properties of two representative lipids.

It is often assumed that the first cells appeared only after the development of a primitive form of protein synthesis. However, it is by no means certain that cells cannot exist without proteins, and it has been suggested that the first cells contained only RNA catalysts. In either case, protein molecules, with their chemically varied side chains, are more powerful catalysts than RNA molecules; therefore, as time passed, cells arose in which RNA served primarily as genetic material, being directly replicated in each generation and inherited by all progeny cells in order to specify proteins.

Both stearic acid (a fatty acid) and phosphatidylcholine (a phospholipid) are composed of chemical groups that form polar "heads" and nonpolar "tails." The polar heads are hydrophilic, or soluble in water, whereas the nonpolar tails are hydrophobic, or insoluble in water. Lipid molecules of this composition spontaneously form aggregate structures such as micelles and lipid bilayers, with their hydrophilic ends oriented toward the watery medium and their hydrophobic ends shielded from the water.

As cells became more complex, a need would have arisen for a stabler form of genetic information storage than that provided by RNA. DNA, related to RNA yet chemically stabler, probably appeared rather late in the evolutionary history of cells. Over a period of time, the genetic information in RNA sequences was transferred to DNA sequences, and the ability of RNA molecules to replicate directly was lost. It was only at this point that the central process of biology—the synthesis, one after the other, of DNA, RNA, and protein—appeared.

Development of Metabolism

The first cells presumably resembled prokaryotic cells in lacking nuclei and functional internal compartments, or organelles. These early cells were also anaerobic (not requiring oxygen), deriving their energy from the fermentation of organic molecules that had previously accumulated on the Earth over long periods of time. Eventually, more sophisticated cells evolved that could carry out primitive forms of photosynthesis, in which light energy was harnessed by membrane-bound proteins to form organic molecules with energy-rich chemical bonds. A major turning point in the evolution of life was the development of photosynthesizing prokaryotes requiring only water as an electron donor and capable of producing molecular oxygen. The descendants of these prokaryotes, the blue-green algae (cyanobacteria), still exist as viable life-forms. Their ancestors prospered to such an extent that the atmosphere became rich in the oxygen they produced. The free availability of this oxygen in turn enabled other prokaryotes to evolve aerobic forms of metabolism that were much more efficient in the use of organic molecules as a source of food.

The switch to predominantly aerobic metabolism is thought to have occurred in bacteria approximately 2 billion years ago, about 1.5 billion years after the first cells had formed. Aerobic eukaryotic cells (cells containing nuclei and all the other organelles) probably appeared about 1.5 billion years ago, their lineage having branched off much earlier from that of the prokaryotes. Eukaryotic cells almost certainly became aerobic by engulfing aerobic prokaryotes, with which they lived in a symbiotic relationship. The mitochondria found in both animals and plants are the descendants of such prokaryotes. Later, in branches of the eukaryotic lineage leading to plants and algae, a blue-green algaelike organism was engulfed to perform photosynthesis. It is likely that over a long period of time these organisms became the chloroplasts.

The eukaryotic cell thus apparently arose as an amalgam of different cells, in the process becoming an efficient aerobic cell whose plasma membrane was freed from energy metabolism—one of the major functions of the cell membrane of prokaryotes. The eukaryotic cell membrane was therefore able to become specialized for cell-to-cell communication and cell signaling. It may be partly for this reason that eukaryotic cells were eventually more successful at forming complex multicellular organisms than their simpler prokaryotic relatives.

Biomolecules

Biomolecules are the macromolecules and ions that are present in organisms. Macromolecules such as lipids, nucleic acid, proteins, etc. are essential for various biological processes such as cell morphogenesis, cell division and development. This chapter has been carefully written to provide an easy understanding of these biomolecules.

A biomolecule refers to any molecule that is produced by living organisms. As such, most of them are organic molecules. The four major groups of biomolecules include polysaccharides, amino acids and proteins, nucleic acids (DNA and RNA), and lipids found in and produced by living organisms. Thus, many of the biomolecules are polymers. A polymer is a compound made up of several repeating units (monomers) or protomers and produced by polymerization. Most of these biomolecules are organic compounds. Being "organic", it means that, in general, they contain carbon atoms covalently bound to other atoms, especially Carbon-Carbon (C-C) and Carbon-Hydrogen (C-H). The four major element constituents are carbon, hydrogen, oxygen, and nitrogen.

The living matter is composed of mainly six elements — carbon, hydrogen, oxygen, nitrogen, phosphorus and sulfur. These elements together constitute about 90% of the dry weight of the human body. Several other functionally important elements are also found in the cells. These include Ca, K, Na, Cl, Mg, Fe, Cu, Co, I, Zn, F, Mo and Se.

Carbon: A Unique Element of Life

Carbon is the most predominant and versatile element of life. It possesses a unique property to form infinite number of compounds. This is attributed to the ability of carbon to form stable covalent bonds and C—C chains of unlimited length. It is estimated that about 90% of compounds found in living system invariably contain carbon.

Chemical Molecules of Life

Life is composed of lifeless chemical molecules. A single cell of the bacterium, Escherichia coli contains about 6,000 different organic compounds. It is believed that man may contain about 100,000 different types of molecules although only a few of them have been characterized.

Table: The major complex biomolecules of cells.

Biomolecule	Building block (repeating unit)	Major functions
Protein	Amino acids	Fundamental basis of structure and function of cell (static and dynamic functions).
Deoxyribonucleic acid (DNA)	Deoxyribonucieotides	Repository of hereditary information.

Ribonucleic acid (RNA)	Ribonudeotides	Essentially required for protein biosynthesis.
Polysaccharide (glycogen)	Monosaccharides (glucose)	Storage form of energy to meet short term demands.
Lipids	Fatty adds, glycerol	Storage form of energy to meet long term demands; structural components of membranes.

Complex Biomolecules

The organic compounds such as amino acids, nucleotides and monosaccharide's serve as the monomeric units or building blocks of complex biomolecules — proteins, nucleic acids (DNA and RNA) and polysaccharides, respectively. The important biomolecules (macromolecules) with their respective building blocks and major functions are given in table. As regards lipids, it may be noted that they are not biopolymers in a strict sense, but majority of them contain fatty acids.

Structural Hierarchy of an Organism

The macromolecules (proteins, lipids, nucleic acids and polysaccharides) form supra-molecular assemblies (e.g. membranes) which in turn organize into organelles, cells, tissues, organs and finally the whole organism.

PROTEIN

Proteins are large biomolecules, or macromolecules, consisting of one or more long chains of amino acid residues. Proteins perform a vast array of functions within organisms, including catalysing metabolic reactions, DNA replication, responding to stimuli, providing structure to cells and organisms, and transporting molecules from one location to another. Proteins differ from one another primarily in their sequence of amino acids, which is dictated by the nucleotide sequence of their genes, and which usually results in protein folding into a specific three-dimensional structure that determines its activity.

A linear chain of amino acid residues is called a polypeptide. A protein contains at least one long polypeptide. Short polypeptides, containing less than 20–30 residues, are rarely considered to be proteins and are commonly called peptides, or sometimes oligopeptides. The individual amino acid residues are bonded together by peptide bonds and adjacent amino acid residues. The sequence of amino acid residues in a protein is defined by the sequence of a gene, which is encoded in the genetic code. In general, the genetic code specifies 20 standard amino acids; however, in certain organisms the genetic code can include selenocysteine and—in certain archaea—pyrrolysine. Shortly after or even during synthesis, the residues in a protein are often chemically modified by post-translational modification, which alters the physical and chemical properties, folding, stability, activity, and ultimately, the function of the proteins. Sometimes proteins have non-peptide groups attached, which can be called prosthetic groups or cofactors. Proteins can also work together to achieve a particular function, and they often associate to form stable protein complexes.

Once formed, proteins only exist for a certain period and are then degraded and recycled by the cell's machinery through the process of protein turnover. A protein's lifespan is measured in terms of its half-life and covers a wide range. They can exist for minutes or years with an average lifespan of 1–2 days in mammalian cells. Abnormal or misfolded proteins are degraded more rapidly either due to being targeted for destruction or due to being unstable.

Like other biological macromolecules such as polysaccharides and nucleic acids, proteins are essential parts of organisms and participate in virtually every process within cells. Many proteins are enzymes that catalyse biochemical reactions and are vital to metabolism. Proteins also have structural or mechanical functions, such as actin and myosin in muscle and the proteins in the cytoskeleton, which form a system of scaffolding that maintains cell shape. Other proteins are important in cell signaling, immune responses, cell adhesion, and the cell cycle. In animals, proteins are needed in the diet to provide the essential amino acids that cannot be synthesized. Digestion breaks the proteins down for use in the metabolism.

Proteins may be purified from other cellular components using a variety of techniques such as ultracentrifugation, precipitation, electrophoresis, and chromatography; the advent of genetic engineering has made possible a number of methods to facilitate purification. Methods commonly used to study protein structure and function include immunohistochemistry, site-directed mutagenesis, X-ray crystallography, nuclear magnetic resonance and mass spectrometry.

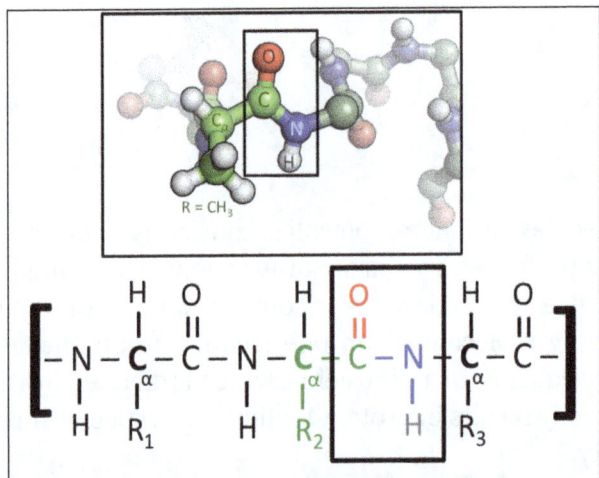

Chemical structure of the peptide bond (bottom) and the three-dimensional structure of a peptide bond between an alanine and an adjacent amino acid (top/inset). The bond itself is made of the CHON elements.

Resonance structures of the peptide bond that links individual amino acids to form a protein polymer

Most proteins consist of linear polymers built from series of up to 20 different L-α- amino acids. All proteinogenic amino acids possess common structural features, including an α-carbon to which an amino group, a carboxyl group, and a variable side chain are bonded. Only proline differs from this basic structure as it contains an unusual ring to the N-end amine group, which forces the CO–NH amide moiety into a fixed conformation. The side chains of the standard amino acids, detailed in the list of standard amino acids, have a great variety of chemical structures and properties; it is the combined effect of all of the amino acid side chains in a protein that ultimately determines its three-dimensional structure and its chemical reactivity. The amino acids in a polypeptide chain are linked by peptide bonds. Once linked in the protein chain, an individual amino acid is called a *residue,* and the linked series of carbon, nitrogen, and oxygen atoms are known as the main chain or protein backbone.

The peptide bond has two resonance forms that contribute some double-bond character and inhibit rotation around its axis, so that the alpha carbons are roughly coplanar. The other two dihedral angles in the peptide bond determine the local shape assumed by the protein backbone. The end with a free amino group is known as the N-terminus or amino terminus, whereas the end of the protein with a free carboxyl group is known as the C-terminus or carboxy terminus (the sequence of the protein is written from N-terminus to C-terminus, from left to right).

The words *protein, polypeptide,* and *peptide* are a little ambiguous and can overlap in meaning. *Protein* is generally used to refer to the complete biological molecule in a stable conformation, whereas *peptide* is generally reserved for a short amino acid oligomers often lacking a stable three-dimensional structure. However, the boundary between the two is not well defined and usually lies near 20–30 residues. *Polypeptide* can refer to any single linear chain of amino acids, usually regardless of length, but often implies an absence of a defined conformation.

Interactions

Proteins can interact with many types of molecules, including with other proteins, with lipids, with carboyhydrates, and with DNA.

Abundance in Cells

It has been estimated that average-sized bacteria contain about 2 million proteins per cell (e.g. *E. coli* and *Staphylococcus aureus*). Smaller bacteria, such as *Mycoplasma* or *spirochetes* contain fewer molecules, on the order of 50,000 to 1 million. By contrast, eukaryotic cells are larger and thus contain much more protein. For instance, yeast cells have been estimated to contain about 50 million proteins and human cells on the order of 1 to 3 billion. The concentration of individual protein copies ranges from a few molecules per cell up to 20 million. Not all genes coding proteins are expressed in most cells and their number depends on, for example, cell type and external stimuli. For instance, of the 20,000 or so proteins encoded by the human genome, only 6,000 are detected in lymphoblastoid cells. Moreover, the number of proteins the genome encodes correlates well with the organism complexity. Eukaryotes have 15,000, bacteria have 3,200, archaea have 2,400, and viruses have 42 proteins on average coded in their respective genomes.

Synthesis

Biosynthesis

Proteins are assembled from amino acids using information encoded in genes. Each protein has its own unique amino acid sequence that is specified by the nucleotide sequence of the gene encoding this protein. The genetic code is a set of three-nucleotide sets called codons and each three-nucleotide combination designates an amino acid, for example AUG (adenine-uracil-guanine) is the code for methionine. Because DNA contains four nucleotides, the total number of possible codons is

64; hence, there is some redundancy in the genetic code, with some amino acids specified by more than one codon. Genes encoded in DNA are first transcribed into pre-messenger RNA (mRNA) by proteins such as RNA polymerase. Most organisms then process the pre-mRNA (also known as a *primary transcript*) using various forms of Post-transcriptional modification to form the mature mRNA, which is then used as a template for protein synthesis by the ribosome. In prokaryotes the mRNA may either be used as soon as it is produced, or be bound by a ribosome after having moved away from the nucleoid. In contrast, eukaryotes make mRNA in the cell nucleus and then translocate it across the nuclear membrane into the cytoplasm, where protein synthesis then takes place. The rate of protein synthesis is higher in prokaryotes than eukaryotes and can reach up to 20 amino acids per second.

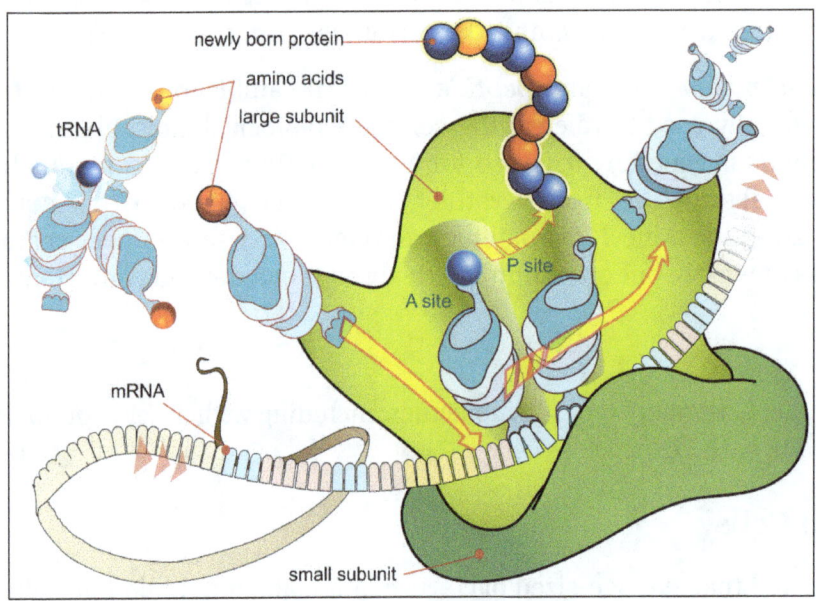

A ribosome produces a protein using mRNA as template.

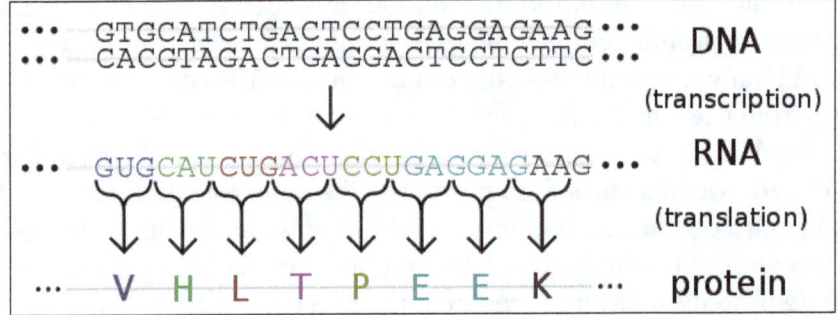

The DNA sequence of a gene encodes the amino acid sequence of a protein.

The process of synthesizing a protein from an mRNA template is known as translation. The mRNA is loaded onto the ribosome and is read three nucleotides at a time by matching each codon to its base pairing anticodon located on a transfer RNA molecule, which carries the amino acid corresponding to the codon it recognizes. The enzyme aminoacyl tRNA synthetase "charges" the tRNA molecules with the correct amino acids. The growing polypeptide is often termed the *nascent chain*. Proteins are always biosynthesized from N-terminus to C-terminus.

The size of a synthesized protein can be measured by the number of amino acids it contains and by its total molecular mass, which is normally reported in units of *daltons* (synonymous with atomic mass units), or the derivative unit kilodalton (kDa). The average size of a protein increases from Archaea to Bacteria to Eukaryote (283, 311, 438 residues and 31, 34, 49 kDa respecitvely) due to a bigger number of protein domains constituting proteins in higher organisms. For instance, yeast proteins are on average 466 amino acids long and 53 kDa in mass. The largest known proteins are the titins, a component of the muscle sarcomere, with a molecular mass of almost 3,000 kDa and a total length of almost 27,000 amino acids.

Chemical Synthesis

Short proteins can also be synthesized chemically by a family of methods known as peptide synthesis, which rely on organic synthesis techniques such as chemical ligation to produce peptides in high yield. Chemical synthesis allows for the introduction of non-natural amino acids into polypeptide chains, such as attachment of fluorescent probes to amino acid side chains. These methods are useful in laboratory biochemistry and cell biology, though generally not for commercial applications. Chemical synthesis is inefficient for polypeptides longer than about 300 amino acids, and the synthesized proteins may not readily assume their native tertiary structure. Most chemical synthesis methods proceed from C-terminus to N-terminus, opposite the biological reaction.

Structure

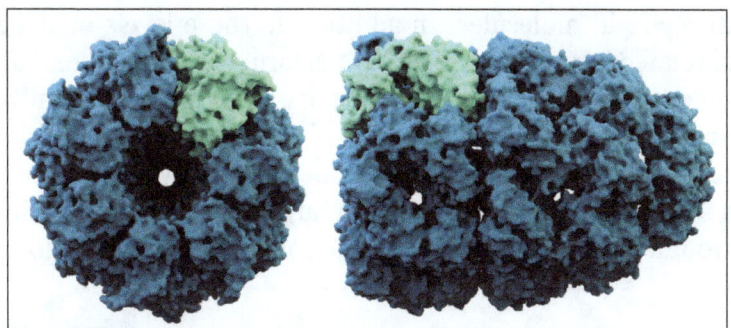

The crystal structure of the chaperonin, a huge protein complex. A single protein subunit is highlighted. Chaperonins assist protein folding.

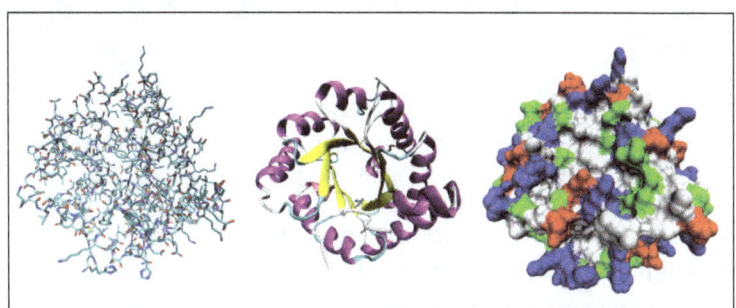

The above figure shows: Three possible representations of the three-dimensional structure of the protein triose phosphate isomerase. Left: All-atom representation colored by atom type. Middle: Simplified representation illustrating the backbone conformation, colored by secondary structure. Right: Solvent-accessible surface representation colored by residue type (acidic residues red, basic residues blue, polar residues green, nonpolar residues white).

Most proteins fold into unique 3-dimensional structures. The shape into which a protein naturally folds is known as its native conformation. Although many proteins can fold unassisted, simply through the chemical properties of their amino acids, others require the aid of molecular chaperones to fold into their native states. Biochemists often refer to four distinct aspects of a protein's structure:

- Primary structure: The amino acid sequence. A protein is a polyamide.

- Secondary structure: Regularly repeating local structures stabilized by hydrogen bonds. The most common examples are the α-helix, β-sheet and turns. Because secondary structures are local, many regions of different secondary structure can be present in the same protein molecule.

- Tertiary structure: The overall shape of a single protein molecule; the spatial relationship of the secondary structures to one another. Tertiary structure is generally stabilized by non-local interactions, most commonly the formation of a hydrophobic core, but also through salt bridges, hydrogen bonds, disulfide bonds, and even posttranslational modifications. The term "tertiary structure" is often used as synonymous with the term fold. The tertiary structure is what controls the basic function of the protein.

- Quaternary structure: The structure formed by several protein molecules (polypeptide chains), usually called protein subunits in this context, which function as a single protein complex.

Proteins are not entirely rigid molecules. In addition to these levels of structure, proteins may shift between several related structures while they perform their functions. In the context of these functional rearrangements, these tertiary or quaternary structures are usually referred to as "conformations", and transitions between them are called *conformational changes*. Such changes are often induced by the binding of a substrate molecule to an enzyme's active site, or the physical region of the protein that participates in chemical catalysis. In solution proteins also undergo variation in structure through thermal vibration and the collision with other molecules.

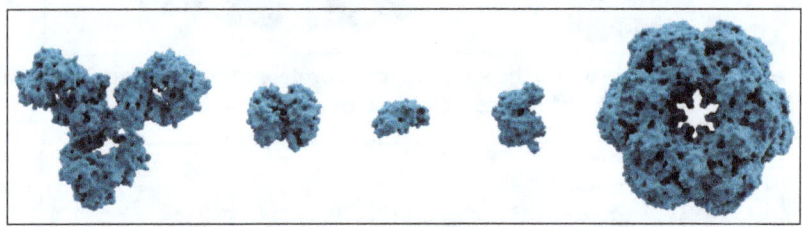

Molecular surface of several proteins showing their comparative sizes. From left to right are: immunoglobulin G (IgG, an antibody), hemoglobin, insulin (a hormone), adenylate kinase (an enzyme), and glutamine synthetase (an enzyme).

Proteins can be informally divided into three main classes, which correlate with typical tertiary structures: globular proteins, fibrous proteins, and membrane proteins. Almost all globular proteins are soluble and many are enzymes. Fibrous proteins are often structural, such as collagen, the major component of connective tissue, or keratin, the protein component of hair and nails. Membrane proteins often serve as receptors or provide channels for polar or charged molecules to pass through the cell membrane.

A special case of intramolecular hydrogen bonds within proteins, poorly shielded from water attack and hence promoting their own dehydration, are called dehydrons.

Protein Domains

Many proteins are composed of several protein domains, i.e. segments of a protein that fold into distinct structural units. Domains usually also have specific functions, such as enzymatic activities (e.g. kinase) or they serve as binding modules (e.g. the SH3 domain binds to proline-rich sequences in other proteins).

Sequence Motif

Short amino acid sequences within proteins often act as recognition sites for other proteins. For instance, SH3 domains typically bind to short PxxP motifs (i.e. 2 prolines [P], separated by 2 unspecified amino acids [x], although the surrounding amino acids may determine the exact binding specificity). A large number of such motifs has been collected in the Eukaryotic Linear Motif (ELM) database.

Cellular Functions

Proteins are the chief actors within the cell, said to be carrying out the duties specified by the information encoded in genes. With the exception of certain types of RNA, most other biological molecules are relatively inert elements upon which proteins act. Proteins make up half the dry weight of an *Escherichia coli* cell, whereas other macromolecules such as DNA and RNA make up only 3% and 20%, respectively. The set of proteins expressed in a particular cell or cell type is known as its proteome.

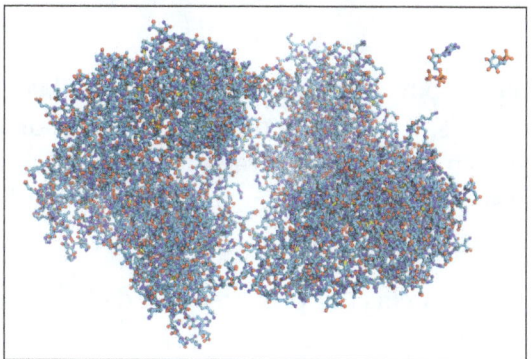

The enzyme hexokinase is shown as a conventional ball-and-stick molecular model.
To scale in the top right-hand corner are two of its substrates, ATP and glucose.

The chief characteristic of proteins that also allows their diverse set of functions is their ability to bind other molecules specifically and tightly. The region of the protein responsible for binding another molecule is known as the binding site and is often a depression or "pocket" on the molecular surface. This binding ability is mediated by the tertiary structure of the protein, which defines the binding site pocket, and by the chemical properties of the surrounding amino acids' side chains. Protein binding can be extraordinarily tight and specific; for example, the ribonuclease inhibitor protein binds to human angiogenin with a sub-femtomolar dissociation constant ($<10^{-15}$ M) but does not bind at all to its amphibian homolog onconase (>1 M). Extremely minor chemical changes such as the addition of a single methyl group to a binding partner can sometimes suffice to nearly eliminate binding; for example, the aminoacyl tRNA synthetase specific to the amino acid valine discriminates against the very similar side chain of the amino acid isoleucine.

Proteins can bind to other proteins as well as to small-molecule substrates. When proteins bind specifically to other copies of the same molecule, they can oligomerize to form fibrils; this process occurs often in structural proteins that consist of globular monomers that self-associate to form rigid fibers. Protein–protein interactions also regulate enzymatic activity, control progression through the cell cycle, and allow the assembly of large protein complexes that carry out many closely related reactions with a common biological function. Proteins can also bind to, or even be integrated into, cell membranes. The ability of binding partners to induce conformational changes in proteins allows the construction of enormously complex signaling networks. As interactions between proteins are reversible, and depend heavily on the availability of different groups of partner proteins to form aggregates that are capable to carry out discrete sets of function, study of the interactions between specific proteins is a key to understand important aspects of cellular function, and ultimately the properties that distinguish particular cell types.

Enzymes

The best-known role of proteins in the cell is as enzymes, which catalyse chemical reactions. Enzymes are usually highly specific and accelerate only one or a few chemical reactions. Enzymes carry out most of the reactions involved in metabolism, as well as manipulating DNA in processes such as DNA replication, DNA repair, and transcription. Some enzymes act on other proteins to add or remove chemical groups in a process known as posttranslational modification. About 4,000 reactions are known to be catalysed by enzymes. The rate acceleration conferred by enzymatic catalysis is often enormous—as much as 10^{17}-fold increase in rate over the uncatalysed reaction in the case of orotate decarboxylase (78 million years without the enzyme, 18 milliseconds with the enzyme).

The molecules bound and acted upon by enzymes are called substrates. Although enzymes can consist of hundreds of amino acids, it is usually only a small fraction of the residues that come in contact with the substrate, and an even smaller fraction—three to four residues on average—that are directly involved in catalysis. The region of the enzyme that binds the substrate and contains the catalytic residues is known as the active site.

Dirigent proteins are members of a class of proteins that dictate the stereochemistry of a compound synthesized by other enzymes.

Cell Signaling and Ligand Binding

Many proteins are involved in the process of cell signaling and signal transduction. Some proteins, such as insulin, are extracellular proteins that transmit a signal from the cell in which they were synthesized to other cells in distant tissues. Others are membrane proteins that act as receptors whose main function is to bind a signaling molecule and induce a biochemical response in the cell. Many receptors have a binding site exposed on the cell surface and an effector domain within the cell, which may have enzymatic activity or may undergo a conformational change detected by other proteins within the cell.

Antibodies are protein components of an adaptive immune system whose main function is to bind antigens, or foreign substances in the body, and target them for destruction. Antibodies can be secreted into the extracellular environment or anchored in the membranes of specialized B cells known as plasma cells. Whereas enzymes are limited in their binding affinity for their substrates by the necessity of conducting their reaction, antibodies have no such constraints. An antibody's binding affinity to its target is extraordinarily high.

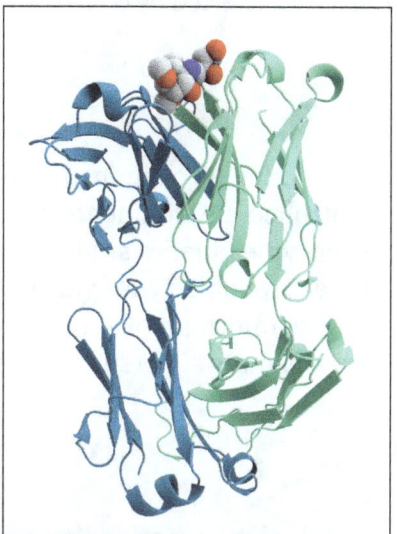

Ribbon diagram of a mouse antibody against cholera that
binds a carbohydrate antigen.

Many ligand transport proteins bind particular small biomolecules and transport them to other locations in the body of a multicellular organism. These proteins must have a high binding affinity when their ligand is present in high concentrations, but must also release the ligand when it is present at low concentrations in the target tissues. The canonical example of a ligand-binding protein is haemoglobin, which transports oxygen from the lungs to other organs and tissues in all vertebrates and has close homologs in every biological kingdom. Lectins are sugar-binding proteins which are highly specific for their sugar moieties. Lectins typically play a role in biological recognition phenomena involving cells and proteins. Receptors and hormones are highly specific binding proteins.

Transmembrane proteins can also serve as ligand transport proteins that alter the permeability of the cell membrane to small molecules and ions. The membrane alone has a hydrophobic core through which polar or charged molecules cannot diffuse. Membrane proteins contain internal channels that allow such molecules to enter and exit the cell. Many ion channel proteins are specialized to select for only a particular ion; for example, potassium and sodium channels often discriminate for only one of the two ions.

Structural Proteins

Structural proteins confer stiffness and rigidity to otherwise-fluid biological components. Most structural proteins are fibrous proteins; for example, collagen and elastin are critical components of connective tissue such as cartilage, and keratin is found in hard or filamentous structures such as hair, nails, feathers, hooves, and some animal shells. Some globular proteins can also play structural functions, for example, actin and tubulin are globular and soluble as monomers, but polymerize to form long, stiff fibers that make up the cytoskeleton, which allows the cell to maintain its shape and size.

Other proteins that serve structural functions are motor proteins such as myosin, kinesin, and dynein, which are capable of generating mechanical forces. These proteins are crucial for cellular motility of single celled organisms and the sperm of many multicellular organisms which reproduce sexually. They also generate the forces exerted by contracting muscles and play essential roles in intracellular transport.

Methods of Study

The activities and structures of proteins may be examined *in vitro, in vivo, and in silico*. In vitro studies of purified proteins in controlled environments are useful for learning how a protein carries out its function: for example, enzyme kinetics studies explore the chemical mechanism of an enzyme's catalytic activity and its relative affinity for various possible substrate molecules. By contrast, *in vivo* experiments can provide information about the physiological role of a protein in the context of a cell or even a whole organism. *In silico* studies use computational methods to study proteins.

Protein Purification

To perform *in vitro* analysis, a protein must be purified away from other cellular components. This process usually begins with cell lysis, in which a cell's membrane is disrupted and its internal contents released into a solution known as a crude lysate. The resulting mixture can be purified using ultracentrifugation, which fractionates the various cellular components into fractions containing soluble proteins; membrane lipids and proteins; cellular organelles, and nucleic acids. Precipitation by a method known as salting out can concentrate the proteins from this lysate. Various types of chromatography are then used to isolate the protein or proteins of interest based on properties such as molecular weight, net charge and binding affinity. The level of purification can be monitored using various types of gel electrophoresis if the desired protein's molecular weight and isoelectric point are known, by spectroscopy if the protein has distinguishable spectroscopic features, or by enzyme assays if the protein has enzymatic activity. Additionally, proteins can be isolated according to their charge using electrofocusing.

For natural proteins, a series of purification steps may be necessary to obtain protein sufficiently pure for laboratory applications. To simplify this process, genetic engineering is often used to add chemical features to proteins that make them easier to purify without affecting their structure or activity. Here, a "tag" consisting of a specific amino acid sequence, often a series of histidine residues (a "His-tag"), is attached to one terminus of the protein. As a result, when the lysate is passed over a chromatography column containing nickel, the histidine residues ligate the nickel and attach to the column while the untagged components of the lysate pass unimpeded. A number of different tags have been developed to help researchers purify specific proteins from complex mixtures.

Cellular Localization

The study of proteins *in vivo* is often concerned with the synthesis and localization of the protein within the cell. Although many intracellular proteins are synthesized in the cytoplasm and membrane-bound or secreted proteins in the endoplasmic reticulum, the specifics of how proteins are targeted to specific organelles or cellular structures is often unclear. A useful technique for assessing cellular localization uses genetic engineering to express in a cell a fusion protein or chimera consisting of the natural protein of interest linked to a "reporter" such as green fluorescent protein (GFP). The fused protein's position within the cell can be cleanly and efficiently visualized using microscopy, as shown in the figure opposite.

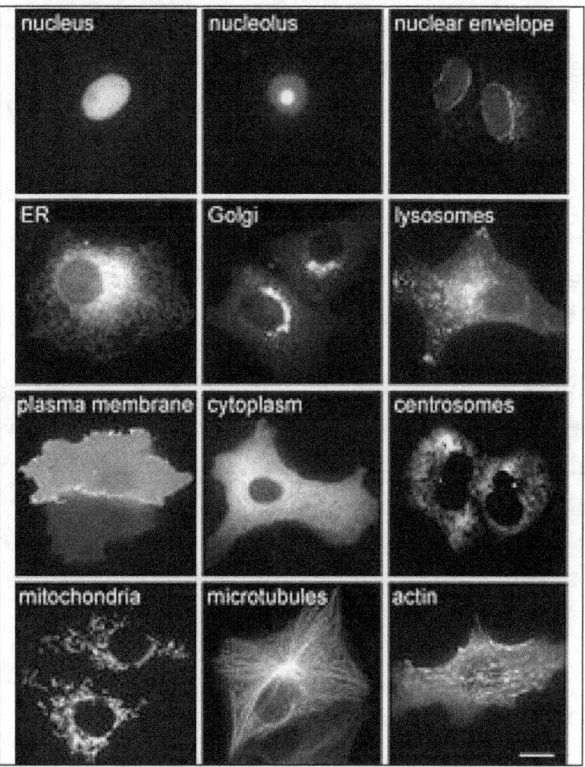

Proteins in different cellular compartments and structures tagged with
green fluorescent protein (here, white).

Other methods for elucidating the cellular location of proteins requires the use of known compartmental markers for regions such as the ER, the Golgi, lysosomes or vacuoles, mitochondria, chloroplasts, plasma membrane, etc. With the use of fluorescently tagged versions of these markers or of antibodies to known markers, it becomes much simpler to identify the localization of a protein of interest. For example, indirect immunofluorescence will allow for fluorescence colocalization and demonstration of location. Fluorescent dyes are used to label cellular compartments for a similar purpose.

Other possibilities exist, as well. For example, immunohistochemistry usually utilizes an antibody to one or more proteins of interest that are conjugated to enzymes yielding either luminescent or chromogenic signals that can be compared between samples, allowing for localization information. Another applicable technique is cofractionation in sucrose (or other material) gradients using isopycnic centrifugation. While this technique does not prove colocalization of a compartment of known density and the protein of interest, it does increase the likelihood, and is more amenable to large-scale studies.

Finally, the gold-standard method of cellular localization is immunoelectron microscopy. This technique also uses an antibody to the protein of interest, along with classical electron microscopy techniques. The sample is prepared for normal electron microscopic examination, and then treated with an antibody to the protein of interest that is conjugated to an extremely electro-dense material, usually gold. This allows for the localization of both ultrastructural details as well as the protein of interest.

Through another genetic engineering application known as site-directed mutagenesis, researchers can alter the protein sequence and hence its structure, cellular localization, and susceptibility to regulation. This technique even allows the incorporation of unnatural amino acids into proteins, using modified tRNAs, and may allow the rational design of new proteins with novel properties.

Proteomics

The total complement of proteins present at a time in a cell or cell type is known as its proteome, and the study of such large-scale data sets defines the field of proteomics, named by analogy to the related field of genomics. Key experimental techniques in proteomics include 2D electrophoresis, which allows the separation of a large number of proteins, mass spectrometry, which allows rapid high-throughput identification of proteins and sequencing of peptides (most often after in-gel digestion), protein microarrays, which allow the detection of the relative levels of a large number of proteins present in a cell, and two-hybrid screening, which allows the systematic exploration of protein–protein interactions. The total complement of biologically possible such interactions is known as the interactome. A systematic attempt to determine the structures of proteins representing every possible fold is known as structural genomics.

Bioinformatics

A vast array of computational methods have been developed to analyze the structure, function, and evolution of proteins.

The development of such tools has been driven by the large amount of genomic and proteomic data available for a variety of organisms, including the human genome. It is simply impossible to study all proteins experimentally, hence only a few are subjected to laboratory experiments while computational tools are used to extrapolate to similar proteins. Such homologous proteins can be efficiently identified in distantly related organisms by sequence alignment. Genome and gene sequences can be searched by a variety of tools for certain properties. Sequence profiling tools can find restriction enzyme sites, open reading frames in nucleotide sequences, and predict secondary structures. Phylogenetic trees can be constructed and evolutionary hypotheses developed using special software like ClustalW regarding the ancestry of modern organisms and the genes they express. The field of bioinformatics is now indispensable for the analysis of genes and proteins.

Structure Determination

Discovering the tertiary structure of a protein, or the quaternary structure of its complexes, can provide important clues about how the protein performs its function and how it can be affected, i.e. in drug design. As proteins are too small to be seen under a light microscope, other methods have to be employed to determine their structure. Common experimental methods include X-ray crystallography and NMR spectroscopy, both of which can produce structural information at atomic resolution. However, NMR experiments are able to provide information from which a subset of distances between pairs of atoms can be estimated, and the final possible conformations for a protein are determined by solving a distance geometry problem. Dual polarisation interferometry is a quantitative analytical method for measuring the overall protein conformation and conformational changes due to interactions or other stimulus. Circular dichroism is another laboratory technique for determining internal β-sheet/α-helical composition of proteins. Cryoelectron microscopy is used to produce lower-resolution structural information about very large protein complexes, including assembled viruses; a variant known as electron crystallography can also produce high-resolution information in some cases, especially for two-dimensional crystals of membrane proteins. Solved structures are usually deposited in the Protein Data Bank (PDB), a freely available resource from which structural data about thousands of proteins can be obtained in the form of Cartesian coordinates for each atom in the protein.

Many more gene sequences are known than protein structures. Further, the set of solved structures is biased toward proteins that can be easily subjected to the conditions required in X-ray crystallography, one of the major structure determination methods. In particular, globular proteins are comparatively easy to crystallize in preparation for X-ray crystallography. Membrane proteins and large protein complexes, by contrast, are difficult to crystallize and are underrepresented in the PDB. Structural genomics initiatives have attempted to remedy these deficiencies by systematically solving representative structures of major fold classes. Protein structure prediction methods attempt to provide a means of generating a plausible structure for proteins whose structures have not been experimentally determined.

Structure Prediction and Simulation

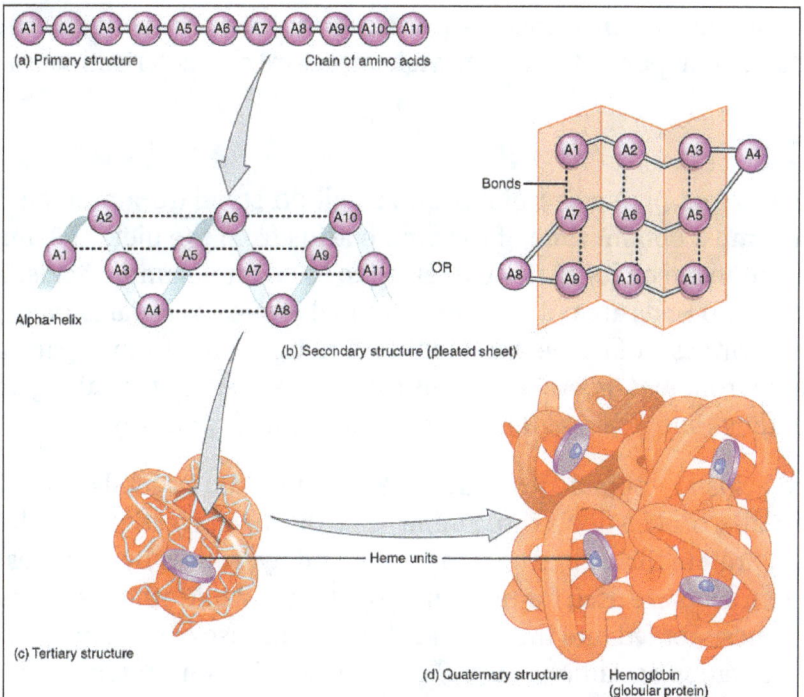

Constituent amino-acids can be analyzed to predict secondary, tertiary and quaternary protein structure, in this case hemoglobin containing heme units.

Complementary to the field of structural genomics, *protein structure prediction* develops efficient mathematical models of proteins to computationally predict the molecular formations in theory, instead of detecting structures with laboratory observation. The most successful type of structure prediction, known as homology modeling, relies on the existence of a "template" structure with sequence similarity to the protein being modeled; structural genomics' goal is to provide sufficient representation in solved structures to model most of those that remain. Although producing accurate models remains a challenge when only distantly related template structures are available, it has been suggested that sequence alignment is the bottleneck in this process, as quite accurate models can be produced if a "perfect" sequence alignment is known. Many structure prediction methods have served to inform the emerging field of protein engineering, in which novel protein folds have already been designed. A more complex computational problem is the prediction of intermolecular interactions, such as in molecular docking and protein–protein interaction prediction.

Mathematical models to simulate dynamic processes of protein folding and binding involve molecular mechanics, in particular, molecular dynamics. Monte Carlo techniques facilitate the computations, which exploit advances in parallel and distributed computing (for example, the Folding@ home project which performs molecular modeling on GPUs). *In silico* simulations discovered the folding of small α-helical protein domains such as the villin headpiece and the HIV accessory protein. Hybrid methods combining standard molecular dynamics with quantum mechanical mathematics explored the electronic states of rhodopsins.

Protein Disorder and Unstructure Prediction

Many proteins (in Eucaryota ~33%) contain large unstructured but biologically functional segments and can be classified as intrinsically disordered proteins. Predicting and analysing protein disorder is, therefore, an important part of protein structure characterisation.

Nutrition

Most microorganisms and plants can biosynthesize all 20 standard amino acids, while animals (including humans) must obtain some of the amino acids from the diet. The amino acids that an organism cannot synthesize on its own are referred to as essential amino acids. Key enzymes that synthesize certain amino acids are not present in animals—such as aspartokinase, which catalyses the first step in the synthesis of lysine, methionine, and threonine from aspartate. If amino acids are present in the environment, microorganisms can conserve energy by taking up the amino acids from their surroundings and downregulating their biosynthetic pathways.

In animals, amino acids are obtained through the consumption of foods containing protein. Ingested proteins are then broken down into amino acids through digestion, which typically involves denaturation of the protein through exposure to acid and hydrolysis by enzymes called proteases. Some ingested amino acids are used for protein biosynthesis, while others are converted to glucose through gluconeogenesis, or fed into the citric acid cycle. This use of protein as a fuel is particularly important under starvation conditions as it allows the body's own proteins to be used to support life, particularly those found in muscle.

In animals such as dogs and cats, protein maintains the health and quality of the skin by promoting hair follicle growth and keratinization, and thus reducing the likelihood of skin problems producing malodours. Poor-quality proteins also have a role regarding gastrointestinal health, increasing the potential for flatulence and odorous compounds in dogs because when proteins reach the colon in an undigested state, they are fermented producing hydrogen sulfide gas, indole, and skatole. Dogs and cats digest animal proteins better than those from plants but products of low-quality animal origin are poorly digested, including skin, feathers, and connective tissue.

NUCLEIC ACID

Nucleic acid is a naturally occurring chemical compound that is capable of being broken down to yield phosphoric acid, sugars, and a mixture of organic bases (purines and pyrimidines). Nucleic acids are the main information-carrying molecules of the cell, and by directing the process of protein synthesis, they determine the inherited characteristics of every living thing. The two main classes of nucleic acids are deoxyribonucleic acid (DNA) and ribonucleic acid (RNA). DNA is the master blueprint for life and constitutes the genetic material in all free-living organisms and most viruses. RNA is the genetic material of certain viruses, but it is also found in all living cells, where it plays an important role in certain processes such as the making of proteins.

Nucleotides: Building Blocks of Nucleic Acids

Basic Structure

Nucleic acids are polynucleotides—that is, long chainlike molecules composed of a series of nearly identical building blocks called nucleotides. Each nucleotide consists of a nitrogen-containing aromatic base attached to a pentose (five-carbon) sugar, which is in turn attached to a phosphate group. Each nucleic acid contains four of five possible nitrogen-containing bases: adenine (A), guanine (G), cytosine (C), thymine (T), and uracil (U). A and G are categorized as purines, and C, T, and U are collectively called pyrimidines. All nucleic acids contain the bases A, C, and G; T, however, is found only in DNA, while U is found in RNA. The pentose sugar in DNA (2'-deoxyribose) differs from the sugar in RNA (ribose) by the absence of a hydroxyl group ($-OH$) on the 2' carbon of the sugar ring. Without an attached phosphate group, the sugar attached to one of the bases is known as a nucleoside. The phosphate group connects successive sugar residues by bridging the 5'-hydroxyl group on one sugar to the 3'-hydroxyl group of the next sugar in the chain. These nucleoside linkages are called phosphodiester bonds and are the same in RNA and DNA.

Biosynthesis and Degradation

Nucleotides are synthesized from readily available precursors in the cell. The ribose phosphate portion of both purine and pyrimidine nucleotides is synthesized from glucose via the pentose phosphate pathway. The six-atom pyrimidine ring is synthesized first and subsequently attached to the ribose phosphate. The two rings in purines are synthesized while attached to the ribose phosphate during the assembly of adenine or guanine nucleosides. In both cases the end product is a nucleotide carrying a phosphate attached to the 5' carbon on the sugar. Finally, a specialized enzyme called a kinase adds two phosphate groups using adenosine triphosphate (ATP) as the phosphate donor to form ribonucleoside triphosphate, the immediate precursor of RNA. For DNA, the 2'-hydroxyl group is removed from the ribonucleoside diphosphate to give deoxyribonucleoside diphosphate. An additional phosphate group from ATP is then added by another kinase to form a deoxyribonucleoside triphosphate, the immediate precursor of DNA.

During normal cell metabolism, RNA is constantly being made and broken down. The purine and pyrimidine residues are reused by several salvage pathways to make more genetic material. Purine is salvaged in the form of the corresponding nucleotide, whereas pyrimidine is salvaged as the nucleoside.

Deoxyribonucleic Acid

Deoxyribonucleic acid, also abbreviated as DNA, is the principal informational macromolecule of the cell, which stores, translates and transfers the genetic information. In the prokaryotes, the DNA is found mostly in the nuclear zone. In eukaryotes it is found in the nucleus, mitochondria and chloroplast. The present understanding of the storage and utilization of the cell's genetic information is based upon the discovery of the structure of DNA by Watson and Crick in 1953.

Structure of DNA

1. DNA is made of two helical chains coiled around the same axis, to form a right-handed double helix.

2. The two chains in the helix are anti-parallel to each other, i.e., the 5′-end of one polynucleotide chain and the 3′-end of the other polynucleotide chain is on the same side and close together.

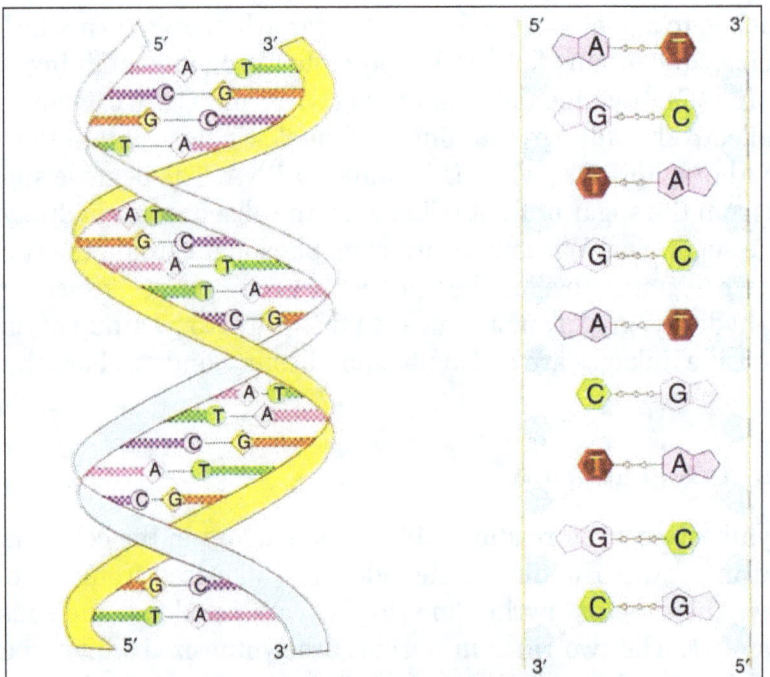

Double helical structure of DNA.

3. The distance between each turn is 3.6 nm (formerly 3.4 nm).

4. There are 10.5 nucleotides per turn (formerly 10 nucleotides).

5. The spatial relationship between the two strands creates major and minor grooves between the two strands. In these grooves some proteins interact.

6. The hydrophilic backbones of alternating deoxyribose and negatively charged phosphate groups are on the outside of the double helix.

7. The hydrophobic pyrimidine and purine bases are inside the double helix, which stabilizes the double helix of the DNA.

8. The double helix is also stabilized by inter-chain hydrogen bond formed between a purine and pyrimidine base.

9. A particular purine base, pairs by hydrogen bonds, only with a particular pyrimidine base, i.e., Adenine (A) pairs with Thymine (T) and Guanine (G) pairs with Cytosine (C) only.

10. Two hydrogen bonds pairs Adenine and Thymine (A = T), whereas three hydrogen bonds pairs Guanine and Cytosine (G ≡ C).

11. The base pairs A = T and G ≡ C are known as complementary base pairs.

12. Due to the presence of complementary base pairing, the two chains of the DNA double helix are complementary to each other.

Hence the number of A' bases are equal to the number of T' bases (or 'G' is equal to 'C) in a given double stranded DNA.

13. One of the strands in the double helix is known as sense strand, i.e., which codes for RNA/ proteins and the other strand is known as antisense strand.

Different Structural Forms of DNA:

The DNA molecules exist in four different structural forms or organizations under different physiological conditions or in different cells or at different points in the same DNA.

Table: Comparison between different structural forms of DNA.

	A	B	Z	H
1.	Shorter and wider	Normal reference strand– Watson and Crick model	Longer and thinner	It is a long stretch (part) of DNA with alternating T and C or polypurine/polypyrimidine
2.	Right-handed double helix	Right-handed double helix	Left-handed double helix	Triple helix
3.	Distance between each turn is 2.3 nm	Distance between each turn is 3.6 nm	Distance between each turn is 3.8 nm	—

| 4. | 11 base pairs per turn | 10.5 base pairs per turn | 12 base pairs per turn | — |
| 5. | Stable in solutions devoid of water | Most stable under physiological conditions | Doubtful existence in physiological state | Helps in gene regulation |

Functions of DNA

The base sequence of the DNA constitutes the informational signal called the genetic material. This nucleotide base sequence enables the DNA to function, store, express and transfer the genetic information. Hence it programs and controls all the activities of an organism directly or indirectly throughout its life cycle.

1. DNA stores the complete genetic information required to specify (form) the structure of all the proteins and RNA's of each organism.

2. DNA is the source of information for the synthesis of all cellular body proteins. Some of the proteins are structural proteins and some are enzymes. These enzymes arrange micro-molecules to form macromolecules. These macromolecules are arranged to form supra-molecular complexes or cell organelles which associate to form cells. These cells group to form tissues which in turn form different organs of a body, specifically peculiar to that organism during foetal development, growth and repair. Hence DNA programs in time and space the orderly biosynthesis of cells and tissue components.

3. It determines the activities of an organism throughout its life cycle, i.e., the period of gestation, birth, maturity, senescence and death.

4. It determines the individuality and identity of a given organism.

5. It duplicates (replicates to form two daughter DNA) itself and transfers one of the copy to the daughter cell during cell division, thus maintaining the genetic material from generation to generation.

Physical Properties of DNA

Denaturation

When DNA is subjected to extremes of pH or temperatures above 80 to 90 degree centigrade, it gets denatured and the double helical structure is unfolded due to disruption of hydrogen bonds between the bases and the hydrophobic interactions of the bases. Finally, the two strands separate completely from each other. This is melting of DNA. The temperature at which a given DNA is denatured to about 50% is known as T_M.

Different DNA melts at different temperatures, which depends upon the $G \equiv C$ content of that DNA. Higher the $G \equiv C$ content, higher is the melting temperature (T_M) and vice-versa. When the temperature or pH is slowly brought back to normal biological range, the two strands will automatically rewind or anneal and will again form the same double helical structure. If the temperature is suddenly cooled down, then the two strands remain separated and exist as single strands.

Denaturation and renaturation of DNA

Buoyant Density

When DNA is centrifuged at high speeds in a concentrated solution of caesium chloride-(CsCI), the CsCl will form a density gradient (ascending) and the DNA will remain stationary or buoyant at a point in the tube where its density is equal to the density of CsCI at that point. Different DNA will have different densities, which again depend upon the GsC content of that DNA. Higher the G = C content, higher is the buoyant density of that DNA and vice versa.

Measurement of these two characters, viz., melting temperature and buoyant density will enable us to calculate the proportions of G $\equiv$ C and A = T pairs in that DNA, which indirectly helps in deducting the gene sequence.

Ribonuclic Acid

RNA is a single-stranded nucleic acid polymer of the four nucleotides A, C, G, and U joined through a backbone of alternating phosphate and ribose sugar residues. It is the first intermediate in converting the information from DNA into proteins essential for the working of a cell. Some RNAs also serve direct roles in cellular metabolism. RNA is made by copying the base sequence of a section of double-stranded DNA, called a gene, into a piece of single-stranded nucleic acid. This process, called transcription, is catalyzed by an enzyme called RNA polymerase.

Chemical Structure

Whereas DNA provides the genetic information for the cell and is inherently quite stable, RNA has many roles and is much more reactive chemically. RNA is sensitive to oxidizing agents such as periodate that lead to opening of the 3'-terminal ribose ring. The 2'-hydroxyl group on the ribose ring is a major cause of instability in RNA, because the presence of alkali leads to rapid cleavage of the phosphodiester bond linking ribose and phosphate groups. In general, this instability is not a significant problem for the cell, because RNA is constantly being synthesized and degraded.

Interactions between the nitrogen-containing bases differ in DNA and RNA. In DNA, which is usually double-stranded, the bases in one strand pair with complementary bases in a second DNA strand. In RNA, which is usually single-stranded, the bases pair with other bases within the same molecule, leading to complex three-dimensional structures. Occasionally, intermolecular RNA/RNA duplexes do form, but they form a right-handed A-type helix rather than the B-type DNA helix. Depending on the amount of salt present, either 11 or 12 base pairs are found in each turn of the helix. Helices between RNA and DNA molecules also form; these adopt the A-type conformation and are more stable than either RNA/RNA or DNA/DNA duplexes. Such hybrid duplexes are important species in biology, being formed when RNA polymerase transcribes DNA into mRNA for protein synthesis and when reverse transcriptase copies a viral RNA genome such as that of the human immunodeficiency virus (HIV).

Single-stranded RNAs are flexible molecules that form a variety of structures through internal base pairing and additional non-base pair interactions. They can form hairpin loops such as those found in transfer RNA (tRNA), as well as longer-range interactions involving both the bases and the phosphate residues of two or more nucleotides. This leads to compact three-dimensional structures. Most of these structures have been inferred from biochemical data, since few crystallographic images are available for RNA molecules. In some types of RNA, a large number of bases are modified after the RNA is transcribed. More than 90 different modifications have been documented, including extensive methylations and a wide variety of substitutions around the ring. In some cases these modifications are known to affect structure and are essential for function.

Types of RNA

Messenger RNA

Messenger RNA (mRNA) delivers the information encoded in one or more genes from the DNA to the ribosome, a specialized structure, or organelle, where that information is decoded into a protein. In prokaryotes, mRNAs contain an exact transcribed copy of the original DNA sequence with a terminal 5'-triphosphate group and a 3'-hydroxyl residue. In eukaryotes the mRNA molecules are more elaborate. The 5'-triphosphate residue is further esterified, forming a structure called a cap. At the 3' ends, eukaryotic mRNAs typically contain long runs of adenosine residues (polyA) that are not encoded in the DNA but are added enzymatically after transcription. Eukaryotic mRNA molecules are usually composed of small segments of the original gene and are generated by a process of cleavage and rejoining from an original precursor RNA (pre-mRNA) molecule, which is an exact copy of the gene. In general, prokaryotic mRNAs are degraded very rapidly, whereas the cap structure and the polyA tail of eukaryotic mRNAs greatly enhance their stability.

Ribosomal RNA

Ribosomal RNA (rRNA) molecules are the structural components of the ribosome. The rRNAs form extensive secondary structures and play an active role in recognizing conserved portions of mRNAs and tRNAs. They also assist with the catalysis of protein synthesis. In the prokaryote E. coli, seven copies of the rRNA genes synthesize about 15,000 ribosomes per cell. In eukaryotes the numbers are much larger. Anywhere from 50 to 5,000 sets of rRNA genes and as many as 10 million ribosomes may be present in a single cell. In eukaryotes these rRNA genes are looped out of the main chromosomal fibres and coalesce in the presence of proteins to form an organelle called the nucleolus. The nucleolus is where the rRNA genes are transcribed and the early assembly of ribosomes takes place.

Transfer RNA

Transfer RNA (tRNA) carries individual amino acids into the ribosome for assembly into the growing polypeptide chain. The tRNA molecules contain 70 to 80 nucleotides and fold into a characteristic cloverleaf structure. Specialized tRNAs exist for each of the 20 amino acids needed for protein synthesis, and in many cases more than one tRNA for each amino acid is present. The nucleotide sequence is converted into a protein sequence by translating each three-base sequence (called a codon) with a specific protein. The 61 codons used to code amino acids can be read by many fewer than 61 distinct tRNAs. In E. coli a total of 40 different tRNAs are used to translate the 61 codons. The amino acids are loaded onto the tRNAs by specialized enzymes called aminoacyl tRNA synthetases, usually with one synthetase for each amino acid. However, in some organisms, less than the full complement of 20 synthetases are required because some amino acids, such as glutamine and asparagine, can be synthesized on their respective tRNAs. All tRNAs adopt similar structures because they all have to interact with the same sites on the ribosome.

Ribozymes

Not all catalysis within the cell is carried out exclusively by proteins. Thomas Cech and Sidney Altman, jointly awarded a Nobel Prize in 1989, discovered that certain RNAs, now known as ribozymes, showed enzymatic activity. Cech showed that a noncoding sequence (intron) in the small subunit rRNA of protozoans, which had to be removed before the rRNA was functional, can excise itself from a much longer precursor RNA molecule and rejoin the two ends in an autocatalytic reaction. Altman showed that the RNA component of an RNA protein complex called ribonuclease P can cleave a precursor tRNA to generate a mature tRNA. In addition to self-splicing RNAs similar to the one discovered by Cech, artificial RNAs have been made that show a variety of catalytic reactions. It is now widely held that there was a stage during evolution when only RNA catalyzed and stored genetic information. This period, sometimes called "the RNA world," is believed to have preceded the function of DNA as genetic material.

Antisense RNAs

Most antisense RNAs are synthetically modified derivatives of RNA or DNA with potential therapeutic value. In nature, antisense RNAs contain sequences that are the complement of the normal coding sequences found in mRNAs (also called sense RNAs). Like mRNAs, antisense RNAs are single-stranded, but they cannot be translated into protein. They can inactivate their complementary mRNA by forming a double-stranded structure that blocks the translation of the base sequence. Artificially introducing antisense RNAs into cells selectively inactivates genes by interfering with normal RNA metabolism.

Viral Genomes

Many viruses use RNA for their genetic material. This is most prevalent among eukaryotic viruses, but a few prokaryotic RNA viruses are also known. Some common examples include poliovirus, human immunodeficiency virus (HIV), and influenza virus, all of which affect humans, and tobacco mosaic virus, which infects plants. In some viruses the entire genetic material is encoded in a single RNA molecule, while in the segmented RNA viruses several RNA molecules may be present. Many RNA viruses such as HIV use a specialized enzyme called reverse transcriptase that permits replication of the virus through a DNA intermediate. In some cases this DNA intermediate becomes integrated into the host chromosome during infection; the virus then exists in a dormant state and effectively evades the host immune system.

Other RNAs

Many other small RNA molecules with specialized functions are present in cells. For example, small nuclear RNAs (snRNAs) are involved in RNA splicing, and other small RNAs that form part of the enzymes telomerase or ribonuclease P are part of ribonucleoprotein particles. The RNA component of telomerase contains a short sequence that serves as a template for the addition of small strings of oligonucleotides at the ends of eukaryotic chromosomes. Other RNA molecules serve as guide RNAs for editing, or they are complementary to small sections of rRNA and either direct the positions at which methyl groups need to be added or mark U residues for conversion to the isomer pseudouridine.

RNA Processing

Cleavage

Following synthesis by transcription, most RNA molecules are processed before reaching their final form. Many rRNA molecules are cleaved from much larger transcripts and may also be methylated or enzymatically modified. In addition, tRNAs are usually formed as longer precursor molecules that are cleaved by ribonuclease P to generate the mature 5′ end and often have extra residues added to their 3′ end to form the sequence CCA. The hydroxyl group on the ribose ring of the terminal A of the 3′-CCA sequence acts as the amino acid acceptor necessary for the function of RNA in protein building.

Splicing

In prokaryotes the protein coding sequence occupies one continuous linear segment of DNA. However, in eukaryotic genes the coding sequences are frequently "split" in the genome—a discovery reached independently in the 1970s by Richard J. Roberts and Phillip A. Sharp, whose work won them a Nobel Prize in 1993. The segments of DNA or RNA coding for protein are called exons, and the noncoding regions separating the exons are called introns. Following transcription, these coding sequences must be joined together before the mRNAs can function. The process of removal of the introns and subsequent rejoining of the exons is called RNA splicing. Each intron is removed in a separate series of reactions by a complicated piece of enzymatic machinery called a spliceosome. This machinery consists of a number of small nuclear ribonucleoprotein particles (snRNPs) that contain small nuclear RNAs (snRNAs).

RNA Editing

Some RNA molecules, particularly those in protozoan mitochondria, undergo extensive editing following their initial synthesis. During this editing process, residues are added or deleted by a post-transcriptional mechanism under the influence of guide RNAs. In some cases as much as 40 percent of the final RNA molecule may be derived by this editing process, rather than being coded directly in the genome. Some examples of editing have also been found in mRNA molecules, but these appear much more limited in scope.

References

- Biomolecule, dictionary: biology-online.org, Retrieved 5 January, 2019

- Gutteridge A, Thornton JM (November 2005). "Understanding nature's catalytic toolkit". Trends in Biochemical Sciences. 30 (11): 622–29. Doi:10.1016/j.tibs.2005.09.006. PMID 16214343

- Biomolecules-top-4-classes-of-biomolecules, biomolecules: biologydiscussion.com, Retrieved 6 February, 2019

- EBI External Services (2010-01-20). "The Catalytic Site Atlas at The European Bioinformatics Institute". Ebi.ac.uk. Retrieved 2011-01-16

- Nucleic-acid, science: britannica.com, Retrieved 7 March, 2019

- Walker JH, Wilson K (2000). Principles and Techniques of Practical Biochemistry. Cambridge, UK: Cambridge University Press. pp. 287–89. ISBN 978-0-521-65873-7

- Na-structure-function-packaging-and-properties-with-diagram, dna: biologydiscussion.com, Retrieved 8 April, 2019

- Methylation, nucleic-acid, science: britannica.com, Retrieved 9 March, 2019

Gene Expression and Regulation

Gene expression is the process through which the genetic code of a gene is used to synthesize the functional gene product. Gene regulation includes the mechanisms that are used by cells to alter the structural and chemical changes in a gene and DNA elements for regulating transcription. This chapter discusses in detail the concepts and processes related to gene expression and regulation.

GENE EXPRESSION

Gene expression is the process by which the genetic code - the nucleotide sequence - of a gene is used to direct protein synthesis and produce the structures of the cell. Genes that code for amino acid sequences are known as 'structural genes'.

The process of gene expression involves two main stages:

- Transcription: the production of messenger RNA (mRNA) by the enzyme RNA polymerase, and the processing of the resulting mRNA molecule.

- Translation: The use of mRNA to direct protein synthesis, and the subsequent post-translational processing of the protein molecule.

Some genes are responsible for the production of other forms of RNA that play a role in translation, including transfer RNA (tRNA) and ribosomal RNA (rRNA).

A structural gene involves a number of different components:

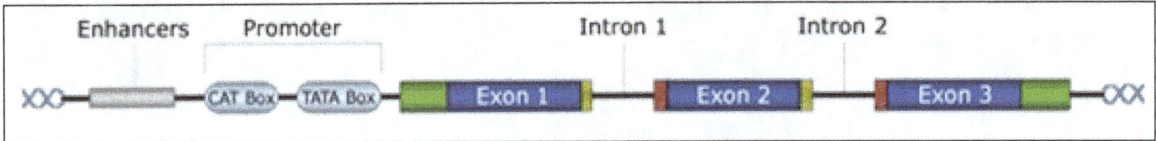

- Exons: Exons code for amino acids and collectively determine the amino acid sequence of the protein product. It is these portions of the gene that are represented in final mature mRNA molecule.

- Introns: Introns are portions of the gene that do not code for amino acids, and are removed (spliced) from the mRNA molecule before translation.

Gene Control Regions

- Start site: A start site for transcription.

- A promoter: A region a few hundred nucleotides 'upstream' of the gene (toward the 5' end). It is not transcribed into mRNA, but plays a role in controlling the transcription of the gene. Transcription factors bind to specific nucleotide sequences in the promoter region and assist in the binding of RNA polymerases.

- Enhancers: Some transcription factors (called activators) bind to regions called 'enhancers' that increase the rate of transcription. These sites may be thousands of nucleotides from the coding sequences or within an intron. Some enhancers are conditional and only work in the presence of other factors as well as transcription factors.

- Silencers: Some transcription factors (called repressors) bind to regions called 'silencers' that depress the rate of transcription.

CENTRAL DOGMA OF MOLECULAR BIOLOGY

The central dogma of molecular biology is an explanation of the flow of genetic information within a biological system. It is often stated as "DNA makes RNA and RNA makes protein," although this is not its original meaning. It was first stated by Francis Crick in 1957, then published in 1958:

> "The Central Dogma. This states that once 'information' has passed into protein it cannot get out again. In more detail, the transfer of information from nucleic acid to nucleic acid, or from nucleic acid to protein may be possible, but transfer from protein to protein, or from protein to nucleic acid is impossible. Information means here the precise determination of sequence, either of bases in the nucleic acid or of amino acid residues in the protein. The central dogma of molecular biology deals with the detailed residue-by-residue transfer of sequential information. It states that such information cannot be transferred back from protein to either protein or nucleic acid."

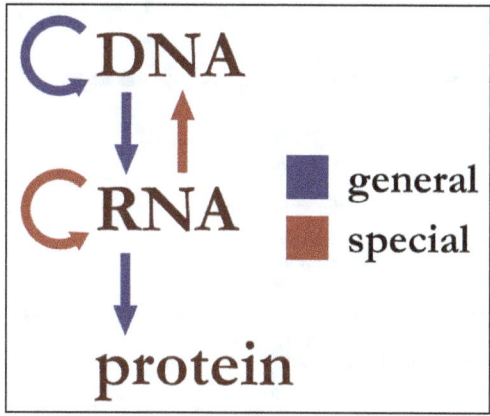

Information flow in biological systems.

A second version of the central dogma is popular but incorrect. This is the simplistic DNA → RNA → protein pathway published by James Watson in the first edition of *The Molecular Biology of the Gene*. Watson's version differs from Crick's because Watson describes a two-step (DNA → RNA and RNA → protein) process as the central dogma. While the dogma, as originally stated by Crick, remains valid today, Watson's version does not.

The dogma is a framework for understanding the transfer of sequence information between information-carrying biopolymers, in the most common or general case, in living organisms. There are 3 major classes of such biopolymers: DNA and RNA (both nucleic acids), and protein. There are 3×3=9 conceivable direct transfers of information that can occur between these. The dogma classes these into 3 groups of 3: three general transfers (believed to occur normally in most cells), three special transfers (known to occur, but only under specific conditions in case of some viruses or in a laboratory), and three unknown transfers (believed never to occur). The general transfers describe the normal flow of biological information: DNA can be copied to DNA (DNA replication), DNA information can be copied into mRNA (transcription), and proteins can be synthesized using the information in mRNA as a template (translation). The special transfers describe: RNA being copied from RNA (RNA replication), DNA being synthesised using an RNA template (reverse transcription), and proteins being synthesised directly from a DNA template without the use of mRNA. The unknown transfers describe: a protein being copied from a protein, synthesis of RNA using the primary structure of a protein as a template, and DNA synthesis using the primary structure of a protein as a template - these are not thought to naturally occur.

Biological Sequence Information

The biopolymers that comprise DNA, RNA and (poly)peptides are linear polymers (i.e.: each monomer is connected to at most two other monomers). The sequence of their monomers effectively encodes information. The transfers of information described by the central dogma ideally are faithful, deterministic transfers, wherein one biopolymer's sequence is used as a template for the construction of another biopolymer with a sequence that is entirely dependent on the original biopolymer's sequence.

General Transfers of Biological Sequential Information

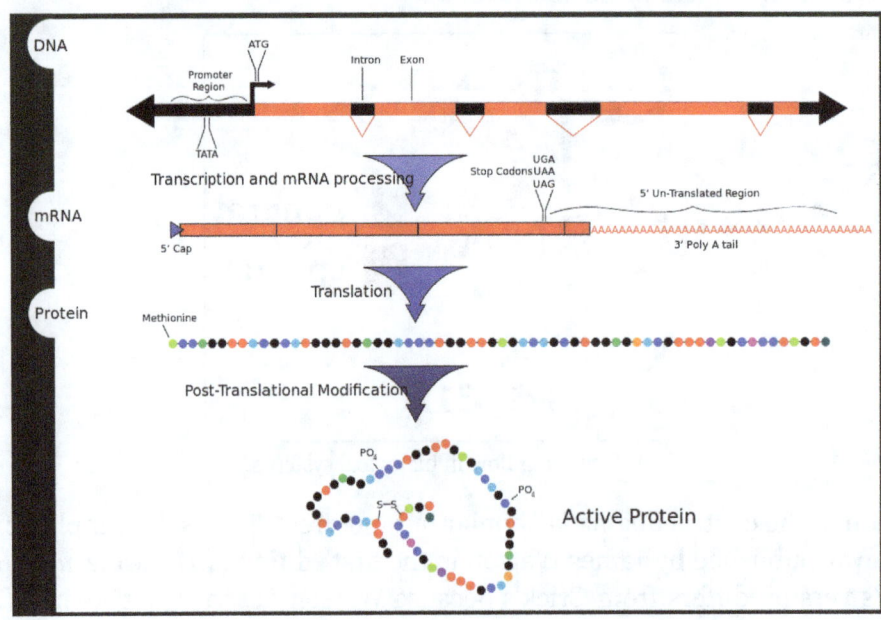

Table of the three classes of information transfer suggested by the dogma.

General	Special	Unknown
DNA → DNA	RNA → DNA	Protein → DNA
DNA → RNA	RNA → RNA	Protein → RNA
RNA → Protein	DNA → Protein	Protein → Protein

DNA Replications

In the sense that DNA replication must occur if genetic material is to be provided for the progeny of any cell, whether somatic or reproductive, the copying from DNA to DNA arguably is the fundamental step in the central dogma. A complex group of proteins called the replisome performs the replication of the information from the parent strand to the complementary daughter strand.

The replisome comprises:

- A helicase that unwinds the superhelix as well as the double-stranded DNA helix to create a replication fork.

- SSB protein that binds open the double-stranded DNA to prevent it from reassociating.

- RNA primase that adds a complementary RNA primer to each template strand as a starting point for replication.

- DNA polymerase III that reads the existing template chain from its 3' end to its 5' end and adds new complementary nucleotides from the 5' end to the 3' end of the daughter chain.

- DNA polymerase I that removes the RNA primers and replaces them with DNA.

- DNA ligase that joins the two Okazaki fragments with phosphodiester bonds to produce a continuous chain.

This process typically takes place during S phase of the cell cycle.

Transcription

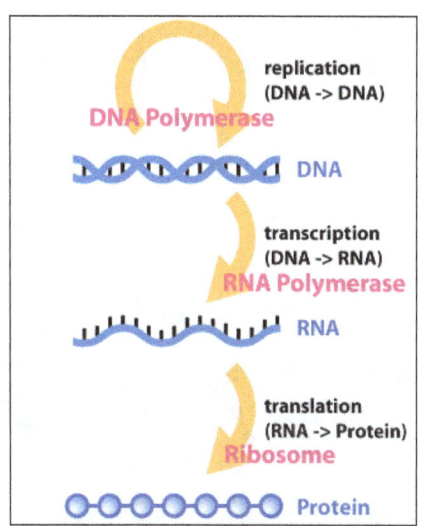

Transcription is the process by which the information contained in a section of DNA is replicated in the form of a newly assembled piece of messenger RNA (mRNA). Enzymes facilitating the process include RNA polymerase and transcription factors. In eukaryotic cells the primary transcript is pre-mRNA. Pre-mRNA must be processed for translation to proceed. Processing includes the addition of a 5' cap and a poly-A tail to the pre-mRNA chain, followed by splicing. Alternative splicing occurs when appropriate, increasing the diversity of the proteins that any single mRNA can produce. The product of the entire transcription process (that began with the production of the pre-mRNA chain) is a mature mRNA chain.

Translation

The mature mRNA finds its way to a ribosome, where it gets translated. In prokaryotic cells, which have no nuclear compartment, the processes of transcription and translation may be linked together without clear separation. In eukaryotic cells, the site of transcription (the cell nucleus) is usually separated from the site of translation (the cytoplasm), so the mRNA must be transported out of the nucleus into the cytoplasm, where it can be bound by ribosomes. The ribosome reads the mRNA triplet codons, usually beginning with an AUG (adenine–uracil–guanine), or initiator methionine codon downstream of the ribosome binding site. Complexes of initiation factors and elongation factors bring aminoacylated transfer RNAs (tRNAs) into the ribosome-mRNA complex, matching the codon in the mRNA to the anti-codon on the tRNA. Each tRNA bears the appropriate amino acid residue to add to the polypeptide chain being synthesised. As the amino acids get linked into the growing peptide chain, the chain begins folding into the correct conformation. Translation ends with a stop codon which may be a UAA, UGA, or UAG triplet.

The mRNA does not contain all the information for specifying the nature of the mature protein. The nascent polypeptide chain released from the ribosome commonly requires additional processing before the final product emerges. For one thing, the correct folding process is complex and vitally important. For most proteins it requires other chaperone proteins to control the form of the product. Some proteins then excise internal segments from their own peptide chains, splicing the free ends that border the gap; in such processes the inside "discarded" sections are called inteins. Other proteins must be split into multiple sections without splicing. Some polypeptide chains need to be cross-linked, and others must be attached to cofactors such as haem (heme) before they become functional.

Special Transfers of Biological Sequential Information

Reverse Transcription

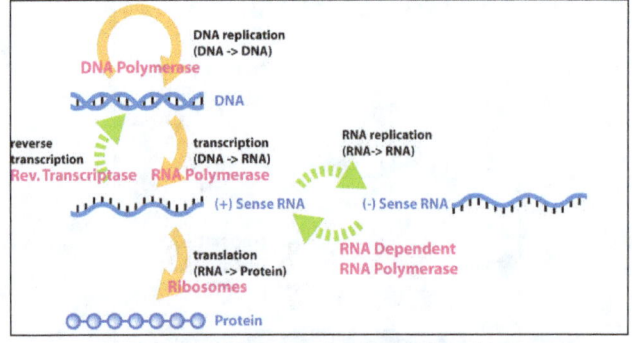

Unusual flows of information highlighted in green.

Reverse transcription is the transfer of information from RNA to DNA (the reverse of normal transcription). This is known to occur in the case of retroviruses, such as HIV, as well as in eukaryotes, in the case of retrotransposons and telomere synthesis. It is the process by which genetic information from RNA gets transcribed into new DNA.

RNA Replication

RNA replication is the copying of one RNA to another. Many viruses replicate this way. The enzymes that copy RNA to new RNA, called RNA-dependent RNA polymerases, are also found in many eukaryotes where they are involved in RNA silencing.

RNA editing, in which an RNA sequence is altered by a complex of proteins and a "guide RNA", could also be seen as an RNA-to-RNA transfer.

Direct Translation from DNA to Protein

Direct translation from DNA to protein has been demonstrated in a cell-free system (i.e. in a test tube), using extracts from *E. coli* that contained ribosomes, but not intact cells. These cell fragments could synthesize proteins from single-stranded DNA templates isolated from other organisms (e,g., mouse or toad), and neomycin was found to enhance this effect. However, it was unclear whether this mechanism of translation corresponded specifically to the genetic code.

Transfers of Information not Explicitly Covered in the Theory

Post-translational Modification

After protein amino acid sequences have been translated from nucleic acid chains, they can be edited by appropriate enzymes. Although this is a form of protein affecting protein sequence, not explicitly covered by the central dogma, there are not many clear examples where the associated concepts of the two fields have much to do with each other.

Inteins

An intein is a "parasitic" segment of a protein that is able to excise itself from the chain of amino acids as they emerge from the ribosome and rejoin the remaining portions with a peptide bond in such a manner that the main protein "backbone" does not fall apart. This is a case of a protein changing its own primary sequence from the sequence originally encoded by the DNA of a gene. Additionally, most inteins contain a homing endonuclease or HEG domain which is capable of finding a copy of the parent gene that does not include the intein nucleotide sequence. On contact with the intein-free copy, the HEG domain initiates the DNA double-stranded break repair mechanism. This process causes the intein sequence to be copied from the original source gene to the intein-free gene. This is an example of protein directly editing DNA sequence, as well as increasing the sequence's heritable propagation.

Methylation

Variation in methylation states of DNA can alter gene expression levels significantly. Methylation variation usually occurs through the action of DNA methylases. When the change is heritable, it is considered epigenetic. When the change in information status is not heritable, it would be a somatic epitype. The effective information content has been changed by means of the actions of a protein or proteins on DNA, but the primary DNA sequence is not altered.

Prions

Prions are proteins of particular amino acid sequences in particular conformations. They propagate themselves in host cells by making conformational changes in other molecules of protein with the same amino acid sequence, but with a different conformation that is functionally important or detrimental to the organism. Once the protein has been transconformed to the prion folding it changes function. In turn it can convey information into new cells and reconfigure more functional molecules of that sequence into the alternate prion form. In some types of prion in fungi this change is continuous and direct; the information flow is Protein → Protein.

Some scientists such as Alain E. Bussard and Eugene Koonin have argued that prion-mediated inheritance violates the central dogma of molecular biology. However, Rosalind Ridley in *Molecular Pathology of the Prions* has written that "The prion hypothesis is not heretical to the central dogma of molecular biology—that the information necessary to manufacture proteins is encoded in the nucleotide sequence of nucleic acid—because it does not claim that proteins replicate. Rather, it claims that there is a source of information within protein molecules that contributes to their biological function, and that this information can be passed on to other molecules."

Natural Genetic Engineering

James A. Shapiro argues that a superset of these examples should be classified as natural genetic engineering and are sufficient to falsify the central dogma. While Shapiro has received a respectful hearing for his view, his critics have not been convinced that his reading of the central dogma is in line with what Crick intended.

DNA REPLICATION

In molecular biology, DNA replication is the biological process of producing two identical replicas of DNA from one original DNA molecule. DNA replication occurs in all living organisms acting as the basis for biological inheritance. The cell possesses the distinctive property of division, which makes replication of DNA essential.

DNA is made up of a double helix of two complementary strands. During replication, these strands are separated. Each strand of the original DNA molecule then serves as a template for the production of its counterpart, a process referred to as semiconservative replication. As a result of semi-conservative replication, the new helix will be composed of an original DNA strand as well as a newly synthesized strand. Cellular proofreading and error-checking mechanisms ensure near perfect fidelity for DNA replication.

In a cell, DNA replication begins at specific locations, or origins of replication, in the genome. Unwinding of DNA at the origin and synthesis of new strands, accommodated by an enzyme known as helicase, results in replication forks growing bi-directionally from the origin. A number of proteins are associated with the replication fork to help in the initiation and continuation of DNA synthesis. Most prominently, DNA polymerase synthesizes the new strands by adding nucleotides that complement each (template) strand. DNA replication occurs during the S-stage of interphase.

DNA replication (DNA amplification) can also be performed *in vitro* (artificially, outside a cell). DNA polymerases isolated from cells and artificial DNA primers can be used to start DNA synthesis at known sequences in a template DNA molecule. Polymerase chain reaction (PCR), ligase chain reaction (LCR), and transcription-mediated amplification (TMA) are examples.

DNA Structure

DNA exists as a double-stranded structure, with both strands coiled together to form the characteristic double-helix. Each single strand of DNA is a chain of four types of nucleotides. Nucleotides in DNA contain a deoxyribose sugar, a phosphate, and a nucleobase. The four types of nucleotide correspond to the four nucleobases adenine, cytosine, guanine, and thymine, commonly abbreviated as A, C, G and T. Adenine and guanine are purine bases, while cytosine and thymine are pyrimidines. These nucleotides form phosphodiester bonds, creating the phosphate-deoxyribose backbone of the DNA double helix with the nucleobases pointing inward (i.e., toward the opposing strand). Nucleobases are matched between strands through hydrogen bonds to form base pairs. Adenine pairs with thymine (two hydrogen bonds), and guanine pairs with cytosine (three hydrogen bonds).

DNA strands have a directionality, and the different ends of a single strand are called the "3′ (three-prime) end" and the "5′ (five-prime) end". By convention, if the base sequence of a single strand of DNA is given, the left end of the sequence is the 5′ end, while the right end of the sequence is the 3′ end. The strands of the double helix are anti-parallel with one being 5′ to 3′, and the opposite strand 3′ to 5′. These terms refer to the carbon atom in deoxyribose to which the next phosphate in the chain attaches. Directionality has consequences in DNA synthesis, because DNA polymerase can synthesize DNA in only one direction by adding nucleotides to the 3′ end of a DNA strand.

The pairing of complementary bases in DNA (through hydrogen bonding) means that the information contained within each strand is redundant. Phosphodiester (intra-strand) bonds are stronger than hydrogen (inter-strand) bonds. This allows the strands to be separated from one another. The nucleotides on a single strand can therefore be used to reconstruct nucleotides on a newly synthesized partner strand.

DNA Polymerase

DNA polymerases are a family of enzymes that carry out all forms of DNA replication. DNA polymerases in general cannot initiate synthesis of new strands, but can only extend an existing DNA or RNA strand paired with a template strand. To begin synthesis, a short fragment of RNA, called a primer, must be created and paired with the template DNA strand.

DNA polymerase adds a new strand of DNA by extending the 3′ end of an existing nucleotide chain, adding new nucleotides matched to the template strand one at a time via the creation of phosphodiester bonds. The energy for this process of DNA polymerization comes from hydrolysis of the high-energy phosphate (phosphoanhydride) bonds between the three phosphates attached to each unincorporated base. Free bases with their attached phosphate groups are called nucleotides; in particular, bases with three attached phosphate groups are called nucleoside triphosphates. When a nucleotide is being added to a growing DNA strand, the formation of a phosphodiester bond between the proximal phosphate of the nucleotide to the growing chain is accompanied by hydrolysis of a high-energy phosphate bond with release of the two distal phosphates as a pyrophosphate. Enzymatic hydrolysis of the resulting pyrophosphate into inorganic phosphate consumes a second high-energy phosphate bond and renders the reaction effectively irreversible.

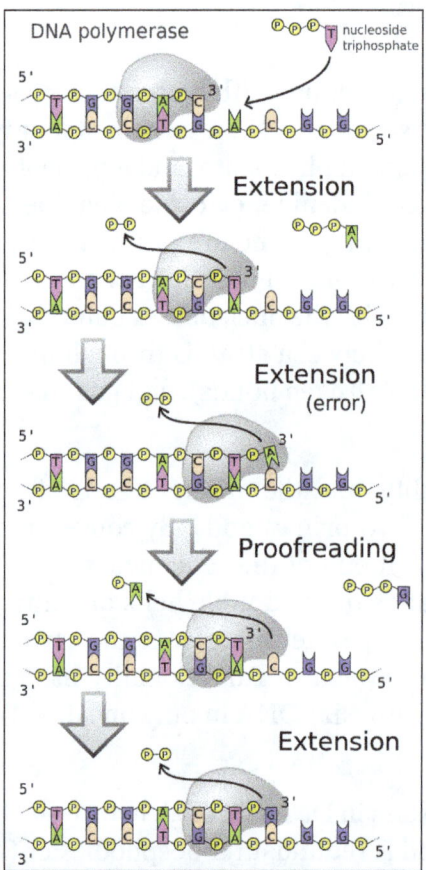

DNA polymerases adds nucleotides to the 3' end of a strand of DNA. If a mismatch is accidentally incorporated, the polymerase is inhibited from further extension. Proofreading removes the mismatched nucleotide and extension continues.

In general, DNA polymerases are highly accurate, with an intrinsic error rate of less than one mistake for every 10^7 nucleotides added. In addition, some DNA polymerases also have proofreading ability; they can remove nucleotides from the end of a growing strand in order to correct mismatched bases. Finally, post-replication mismatch repair mechanisms monitor the DNA for errors, being capable of distinguishing mismatches in the newly synthesized DNA strand from the original strand sequence. Together, these three discrimination steps enable replication fidelity of less than one mistake for every 10^9 nucleotides added.

The rate of DNA replication in a living cell was first measured as the rate of phage T4 DNA elongation in phage-infected *E. coli*. During the period of exponential DNA increase at 37 °C, the rate was 749 nucleotides per second. The mutation rate per base pair per replication during phage T4 DNA synthesis is 1.7 per 10^8.

Replication Process

DNA replication, like all biological polymerization processes, proceeds in three enzymatically catalyzed and coordinated steps: initiation, elongation and termination.

Initiation

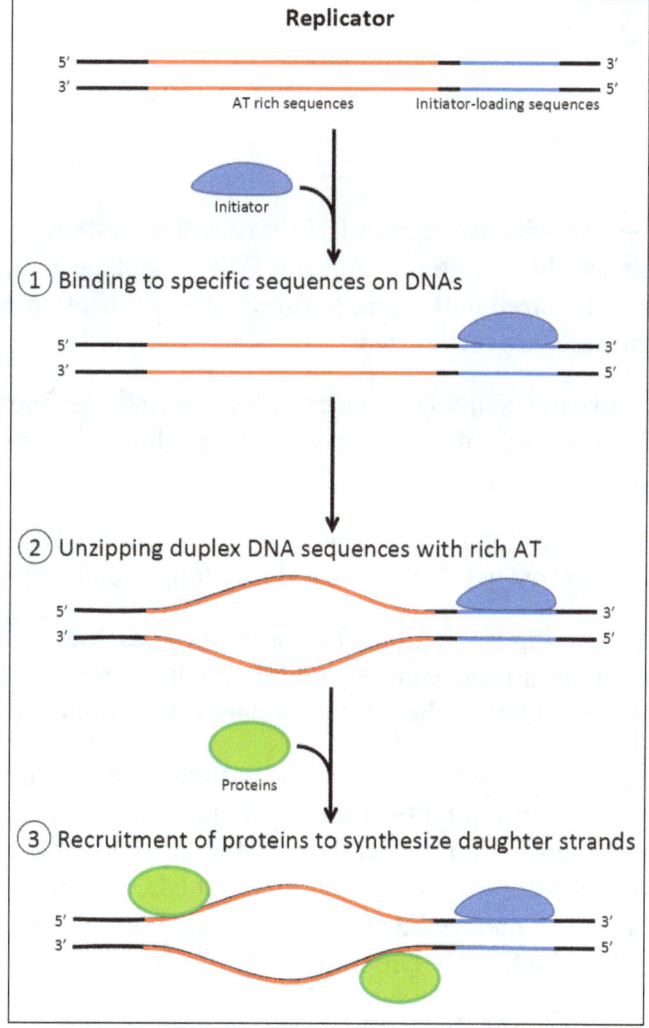

Role of initiators for initiation of DNA replication.

For a cell to divide, it must first replicate its DNA. This process is initiated at particular points in the DNA, known as "origins", which are targeted by initiator proteins. In *E. coli* this protein is DnaA; in yeast, this is the origin recognition complex. Sequences used by initiator proteins tend to be "AT-rich" (rich in adenine and thymine bases), because A-T base pairs have two hydrogen bonds (rather than the three formed in a C-G pair) and thus are easier to strand-separate. Once the origin has been located, these initiators recruit other proteins and form the pre-replication complex, which unwinds the double-stranded DNA.

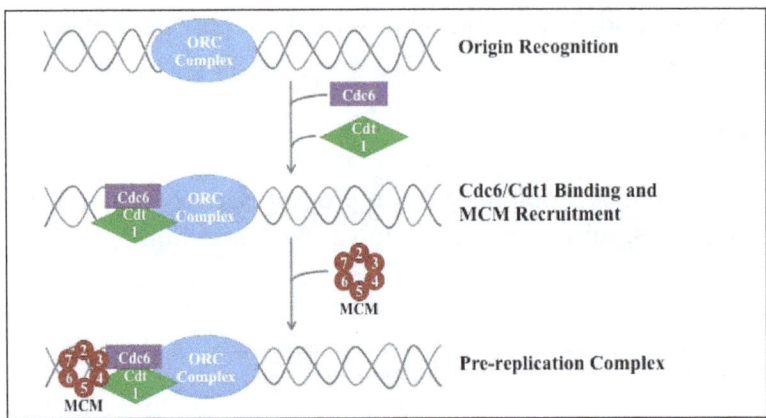

Formation of pre-replication complex.

Elongation

DNA polymerase has 5'–3' activity. All known DNA replication systems require a free 3' hydroxyl group before synthesis can be initiated (note: the DNA template is read in 3' to 5' direction whereas a new strand is synthesized in the 5' to 3' direction—this is often confused). Four distinct mechanisms for DNA synthesis are recognized:

1. All cellular life forms and many DNA viruses, phages and plasmids use a primase to synthesize a short RNA primer with a free 3' OH group which is subsequently elongated by a DNA polymerase.

2. The retroelements (including retroviruses) employ a transfer RNA that primes DNA replication by providing a free 3' OH that is used for elongation by the reverse transcriptase.

3. In the adenoviruses and the φ29 family of bacteriophages, the 3' OH group is provided by the side chain of an amino acid of the genome attached protein (the terminal protein) to which nucleotides are added by the DNA polymerase to form a new strand.

4. In the single stranded DNA viruses—a group that includes the circoviruses, the geminiviruses, the parvoviruses and others—and also the many phages and plasmids that use the rolling circle replication (RCR) mechanism, the RCR endonuclease creates a nick in the genome strand (single stranded viruses) or one of the DNA strands (plasmids). The 5' end of the nicked strand is transferred to a tyrosine residue on the nuclease and the free 3' OH group is then used by the DNA polymerase to synthesize the new strand.

The first is the best known of these mechanisms and is used by the cellular organisms. In this mechanism, once the two strands are separated, primase adds RNA primers to the template strands. The leading strand receives one RNA primer while the lagging strand receives several. The leading strand is continuously extended from the primer by a DNA polymerase with high processivity, while the lagging strand is extended discontinuously from each primer forming Okazaki fragments. RNase removes the primer RNA fragments, and a low processivity DNA polymerase distinct from the replicative polymerase enters to fill the gaps. When this is complete, a single nick on the leading strand and several nicks on the lagging strand can be found. Ligase works to fill these nicks in, thus completing the newly replicated DNA molecule.

The primase used in this process differs significantly between bacteria and archaea/eukaryotes. Bacteria use a primase belonging to the DnaG protein superfamily which contains a catalytic domain of the TOPRIM fold type. The TOPRIM fold contains an α/β core with four conserved strands in a Rossmann-like topology. This structure is also found in the catalytic domains of topoisomerase Ia, topoisomerase II, the OLD-family nucleases and DNA repair proteins related to the RecR protein.

The primase used by archaea and eukaryotes, in contrast, contains a highly derived version of the RNA recognition motif (RRM). This primase is structurally similar to many viral RNA-dependent RNA polymerases, reverse transcriptases, cyclic nucleotide generating cyclases and DNA polymerases of the A/B/Y families that are involved in DNA replication and repair. In eukaryotic replication, the primase forms a complex with Pol α.

Multiple DNA polymerases take on different roles in the DNA replication process. In *E. coli*, DNA Pol III is the polymerase enzyme primarily responsible for DNA replication. It assembles into a replication complex at the replication fork that exhibits extremely high processivity, remaining intact for the entire replication cycle. In contrast, DNA Pol I is the enzyme responsible for replacing RNA primers with DNA. DNA Pol I has a 5′ to 3′ exonuclease activity in addition to its polymerase activity, and uses its exonuclease activity to degrade the RNA primers ahead of it as it extends the DNA strand behind it, in a process called nick translation. Pol I is much less processive than Pol III because its primary function in DNA replication is to create many short DNA regions rather than a few very long regions.

In eukaryotes, the low-processivity enzyme, Pol α, helps to initiate replication because it forms a complex with primase. In eukaryotes, leading strand synthesis is thought to be conducted by Pol ε; however, this view has recently been challenged, suggesting a role for Pol δ. Primer removal is completed Pol δ while repair of DNA during replication is completed by Pol ε.

As DNA synthesis continues, the original DNA strands continue to unwind on each side of the bubble, forming a replication fork with two prongs. In bacteria, which have a single origin of replication on their circular chromosome, this process creates a "theta structure" (resembling the Greek letter theta: θ). In contrast, eukaryotes have longer linear chromosomes and initiate replication at multiple origins within these.

Replication Fork

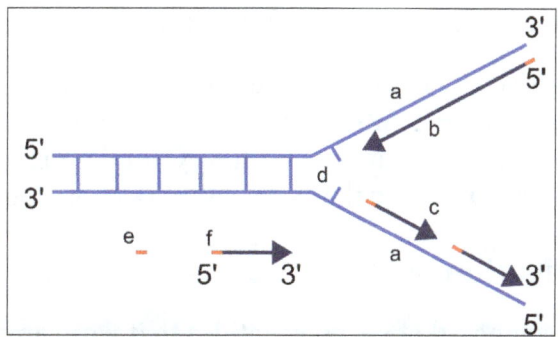

Scheme of the replication fork. a: template, b: leading strand, c: lagging strand, d: replication fork, e: primer, f: Okazaki fragments.

The replication fork is a structure that forms within the long helical DNA during DNA replication. It is created by helicases, which break the hydrogen bonds holding the two DNA strands together in the helix. The resulting structure has two branching "prongs", each one made up of a single strand of DNA. These two strands serve as the template for the leading and lagging strands, which will be created as DNA polymerase matches complementary nucleotides to the templates; the templates may be properly referred to as the leading strand template and the lagging strand template.

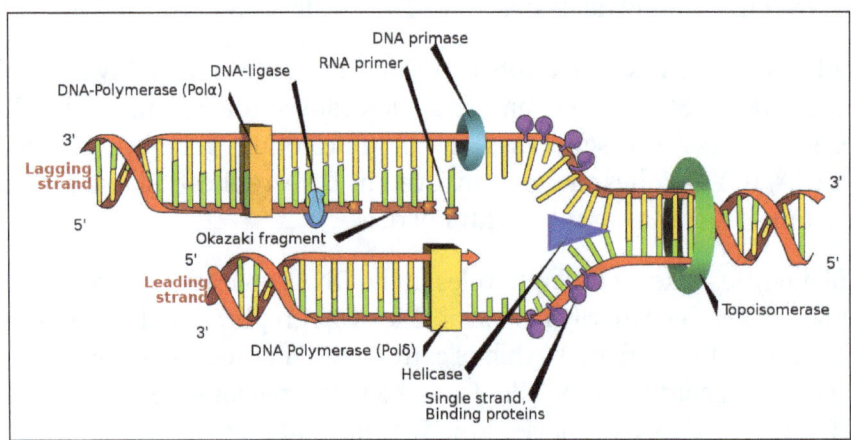

Many enzymes are involved in the DNA replication fork.

DNA is always synthesized in the 5′ to 3′ direction. Since the leading and lagging strand templates are oriented in opposite directions at the replication fork, a major issue is how to achieve synthesis of nascent (new) lagging strand DNA, whose direction of synthesis is opposite to the direction of the growing replication fork.

Leading Strand

The leading strand is the strand of nascent DNA which is synthesized in the same direction as the growing replication fork. This sort of DNA replication is continuous.

Lagging Strand

The lagging strand is the strand of nascent DNA whose direction of synthesis is opposite to the direction of the growing replication fork. Because of its orientation, replication of the lagging strand is more complicated as compared to that of the leading strand. As a consequence, the DNA polymerase on this strand is seen to "lag behind" the other strand.

The lagging strand is synthesized in short, separated segments. On the lagging strand *template*, a primase "reads" the template DNA and initiates synthesis of a short complementary RNA primer. A DNA polymerase extends the primed segments, forming Okazaki fragments. The RNA primers are then removed and replaced with DNA, and the fragments of DNA are joined together by DNA ligase.

Dynamics at the Replication Fork

In all cases the helicase is composed of six polypeptides that wrap around only one strand of the DNA being replicated. The two polymerases are bound to the helicase heximer. In eukaryotes the helicase wraps around the leading strand, and in prokaryotes it wraps around the lagging strand.

As helicase unwinds DNA at the replication fork, the DNA ahead is forced to rotate. This process results in a build-up of twists in the DNA ahead. This build-up forms a torsional resistance that would eventually halt the progress of the replication fork. Topoisomerases are enzymes that temporarily break the strands of DNA, relieving the tension caused by unwinding the two strands of the DNA helix; topoisomerases (including DNA gyrase) achieve this by adding negative supercoils to the DNA helix.

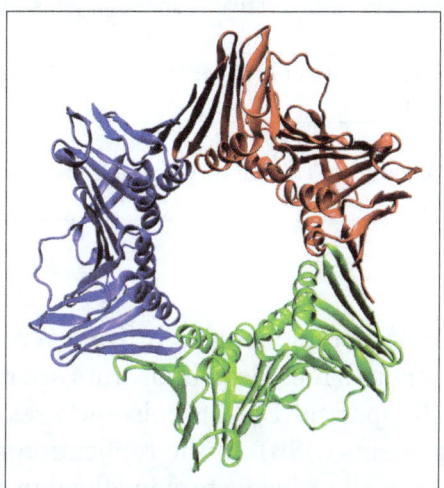

The assembled human DNA clamp, a trimer of the protein PCNA.

Bare single-stranded DNA tends to fold back on itself forming secondary structures; these structures can interfere with the movement of DNA polymerase. To prevent this, single-strand binding proteins bind to the DNA until a second strand is synthesized, preventing secondary structure formation.

Clamp proteins form a sliding clamp around DNA, helping the DNA polymerase maintain contact with its template, thereby assisting with processivity. The inner face of the clamp enables DNA to be threaded through it. Once the polymerase reaches the end of the template or detects double-stranded DNA, the sliding clamp undergoes a conformational change that releases the DNA polymerase. Clamp-loading proteins are used to initially load the clamp, recognizing the junction between template and RNA primers.

DNA Replication Proteins

At the replication fork, many replication enzymes assemble on the DNA into a complex molecular machine called the replisome. The following is a list of major DNA replication enzymes that participate in the replisome:

Enzyme	Function in DNA replication
DNA Helicase	Also known as helix destabilizing enzyme. Helicase separates the two strands of DNA at the Replication Fork behind the topoisomerase.
DNA Polymerase	The enzyme responsible for catalyzing the addition of nucleotide substrates to DNA in the 5′ to 3′ direction during DNA replication. Also performs proof-reading and error correction. There exist many different types of DNA Polymerase, each of which perform different functions in different types of cells.
DNA clamp	A protein which prevents elongating DNA polymerases from dissociating from the DNA parent strand.

Single-Strand Binding (SSB) Proteins	Bind to ssDNA and prevent the DNA double helix from re-annealing after DNA helicase unwinds it, thus maintaining the strand separation, and facilitating the synthesis of the nascent strand.
Topoisomerase	Relaxes the DNA from its super-coiled nature.
DNA Gyrase	Relieves strain of unwinding by DNA helicase; this is a specific type of topoisomerase
DNA Ligase	Re-anneals the semi-conservative strands and joins Okazaki Fragments of the lagging strand.
Primase	Provides a starting point of RNA (or DNA) for DNA polymerase to begin synthesis of the new DNA strand.
Telomerase	Lengthens telomeric DNA by adding repetitive nucleotide sequences to the ends of eukaryotic chromosomes. This allows germ cells and stem cells to avoid the Hayflick limit on cell division.

Replication Machinery

Replication machineries consist of factors involved in DNA replication and appearing on template ssDNAs. Replication machineries include primosotors are replication enzymes; DNA polymerase, DNA helicases, DNA clamps and DNA topoisomerases, and replication proteins; e.g. single-stranded DNA binding proteins (SSB). In the replication machineries these components coordinate. In most of the bacteria, all of the factors involved in DNA replication are located on replication forks and the complexes stay on the forks during DNA replication. These replication machineries are called replisomes or DNA replicase systems. These terms are generic terms for proteins located on replication forks. In eukaryotic and some bacterial cells the replisomes are not formed.

Since replication machineries do not move relatively to template DNAs such as factories, they are called a replication factory. In an alternative figure, DNA factories are similar to projectors and DNAs are like as cinematic films passing constantly into the projectors. In the replication factory model, after both DNA helicases for leading strands and lagging strands are loaded on the template DNAs, the helicases run along the DNAs into each other. The helicases remain associated for the remainder of replication process. Peter Meister et al. observed directly replication sites in budding yeast by monitoring green fluorescent protein(GFP)-tagged DNA polymerases α. They detected DNA replication of pairs of the tagged loci spaced apart symmetrically from a replication origin and found that the distance between the pairs decreased markedly by time. This finding suggests that the mechanism of DNA replication goes with DNA factories. That is, couples of replication factories are loaded on replication origins and the factories associated with each other. Also, template DNAs move into the factories, which bring extrusion of the template ssDNAs and nascent DNAs. Meister's finding is the first direct evidence of replication factory model. Subsequent research has shown that DNA helicases form dimers in many eukaryotic cells and bacterial replication machineries stay in single intranuclear location during DNA synthesis.

The replication factories perform disentanglement of sister chromatids. The disentanglement is essential for distributing the chromatids into daughter cells after DNA replication. Because sister chromatids after DNA replication hold each other by Cohesin rings, there is the only chance for the disentanglement in DNA replication. Fixing of replication machineries as replication factories can improve the success rate of DNA replication. If replication forks move freely in chromosomes, catenation of nuclei is aggravated and impedes mitotic segregation.

Termination

Eukaryotes initiate DNA replication at multiple points in the chromosome, so replication forks meet and terminate at many points in the chromosome. Because eukaryotes have linear chromosomes, DNA replication is unable to reach the very end of the chromosomes. Due to this problem, DNA is lost in each replication cycle from the end of the chromosome. Telomeres are regions of repetitive DNA close to the ends and help prevent loss of genes due to this shortening. Shortening of the telomeres is a normal process in somatic cells. This shortens the telomeres of the daughter DNA chromosome. As a result, cells can only divide a certain number of times before the DNA loss prevents further division. This is known as the Hayflick limit. Within the germ cell line, which passes DNA to the next generation, telomerase extends the repetitive sequences of the telomere region to prevent degradation. Telomerase can become mistakenly active in somatic cells, sometimes leading to cancer formation. Increased telomerase activity is one of the hallmarks of cancer.

Termination requires that the progress of the DNA replication fork must stop or be blocked. Termination at a specific locus, when it occurs, involves the interaction between two components: (1) a termination site sequence in the DNA, and (2) a protein which binds to this sequence to physically stop DNA replication. In various bacterial species, this is named the DNA replication terminus site-binding protein, or Ter protein.

Because bacteria have circular chromosomes, termination of replication occurs when the two replication forks meet each other on the opposite end of the parental chromosome. *E. coli* regulates this process through the use of termination sequences that, when bound by the Tus protein, enable only one direction of replication fork to pass through. As a result, the replication forks are constrained to always meet within the termination region of the chromosome.

Regulation

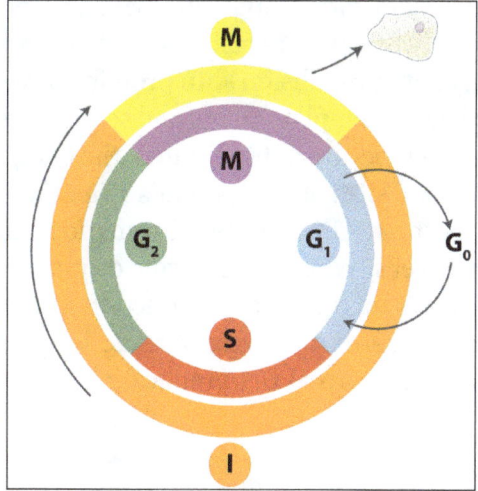

The cell cycle of eukaryotic cells.

Eukaryotes

Within eukaryotes, DNA replication is controlled within the context of the cell cycle. As the cell grows and divides, it progresses through stages in the cell cycle; DNA replication takes place during the S phase (synthesis phase). The progress of the eukaryotic cell through the cycle is controlled by cell cycle checkpoints. Progression through checkpoints is controlled through complex interactions between various proteins, including cyclins and cyclin-dependent kinases. Unlike bacteria, eukaryotic DNA replicates in the confines of the nucleus.

The G1/S checkpoint (or restriction checkpoint) regulates whether eukaryotic cells enter the process of DNA replication and subsequent division. Cells that do not proceed through this checkpoint remain in the G0 stage and do not replicate their DNA.

Replication of chloroplast and mitochondrial genomes occurs independently of the cell cycle, through the process of D-loop replication.

Replication Focus

In vertebrate cells, replication sites concentrate into positions called replication foci. Replication sites can be detected by immunostaining daughter strands and replication enzymes and monitoring GFP-tagged replication factors. By these methods it is found that replication foci of varying size and positions appear in S phase of cell division and their number per nucleus is far smaller than the number of genomic replication forks.

P. Heun et al. tracked GFP-tagged replication foci in budding yeast cells and revealed that replication origins move constantly in G1 and S phase and the dynamics decreased significantly in S phase. Traditionally, replication sites were fixed on spatial structure of chromosomes by nuclear matrix or lamins. The Heun's results denied the traditional concepts, budding yeasts don't have lamins, and support that replication origins self-assemble and form replication foci.

By firing of replication origins, controlled spatially and temporally, the formation of replication foci is regulated. D. A. Jackson et al. revealed that neighboring origins fire simultaneously in mammalian cells. Spatial juxtaposition of replication sites brings clustering of replication forks. The clustering do rescue of stalled replication forks and favors normal progress of replication forks. Progress of replication forks is inhibited by many factors; collision with proteins or with complexes binding strongly on DNA, deficiency of dNTPs, nicks on template DNAs and so on. If replication forks stall and the remaining sequences from the stalled forks are not replicated, the daughter strands have nick obtained un-replicated sites. The un-replicated sites on one parent's strand hold the other strand together but not daughter strands. Therefore, the resulting sister chromatids cannot separate from each other and cannot divide into 2 daughter cells. When neighboring origins fire and a fork from one origin is stalled, fork from other origin access on an opposite direction of the stalled fork and duplicate the un-replicated sites. As other mechanism of the rescue there is application of dormant replication origins that excess origins don't fire in normal DNA replication.

Bacteria

Most bacteria do not go through a well-defined cell cycle but instead continuously copy their DNA; during rapid growth, this can result in the concurrent occurrence of multiple rounds of replication. In *E. coli*, the best-characterized bacteria, DNA replication is regulated through several mechanisms, including: the hemimethylation and sequestering of the origin sequence, the ratio of adenosine triphosphate (ATP) to adenosine diphosphate (ADP), and the levels of protein DnaA. All these control the binding of initiator proteins to the origin sequences.

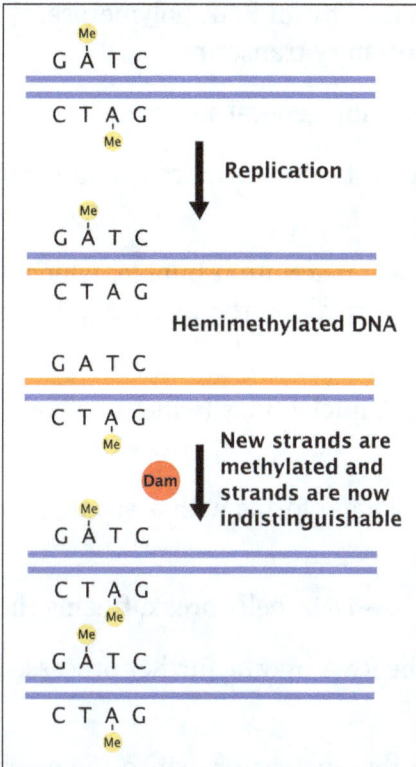

Dam methylates adenine of GATC sites after replication.

Because *E. coli* methylates GATC DNA sequences, DNA synthesis results in hemimethylated sequences. This hemimethylated DNA is recognized by the protein SeqA, which binds and sequesters the origin sequence; in addition, DnaA (required for initiation of replication) binds less well to hemimethylated DNA. As a result, newly replicated origins are prevented from immediately initiating another round of DNA replication.

ATP builds up when the cell is in a rich medium, triggering DNA replication once the cell has reached a specific size. ATP competes with ADP to bind to DnaA, and the DnaA-ATP complex is able to initiate replication. A certain number of DnaA proteins are also required for DNA replication — each time the origin is copied, the number of binding sites for DnaA doubles, requiring the synthesis of more DnaA to enable another initiation of replication.

In fast-growing bacteria, such as *E. coli*, chromosome replication takes more time than dividing the cell. The bacteria solve this by initiating a new round of replication before the previous one has been terminated. The new round of replication will form the chromosome of the cell that is born two generations after the dividing cell. This mechanism creates overlapping replication cycles.

TRANSCRIPTION

Transcription is the first step of DNA based gene expression, in which a particular segment of DNA is copied into RNA (especially mRNA) by the enzyme RNA polymerase. Both DNA and RNA are nucleic acids, which use base pairs of nucleotides as a complementary language. During transcription, a DNA sequence is read by an RNA polymerase, which produces a complementary, antiparallel RNA strand called a primary transcript.

Transcription proceeds in the following general steps:

1. RNA polymerase, together with one or more general transcription factors, binds to promoter DNA.

2. RNA polymerase creates a transcription bubble, which separates the two strands of the DNA helix. This is done by breaking the hydrogen bonds between complementary DNA nucleotides.

3. RNA polymerase adds RNA nucleotides (which are complementary to the nucleotides of one DNA strand).

4. RNA sugar-phosphate backbone forms with assistance from RNA polymerase to form an RNA strand.

5. Hydrogen bonds of the RNA–DNA helix break, freeing the newly synthesized RNA strand.

6. If the cell has a nucleus, the RNA may be further processed. This may include polyadenylation, capping, and splicing.

7. The RNA may remain in the nucleus or exit to the cytoplasm through the nuclear pore complex.

The stretch of DNA transcribed into an RNA molecule is called a *transcription unit* and encodes at least one gene. If the gene encodes a protein, the transcription produces messenger RNA (mRNA); the mRNA, in turn, serves as a template for the protein's synthesis through translation. Alternatively, the transcribed gene may encode for non-coding RNA such as microRNA, ribosomal RNA (rRNA), transfer RNA (tRNA), or enzymatic RNA molecules called ribozymes. Overall, RNA helps synthesize, regulate, and process proteins; it therefore plays a fundamental role in performing functions within a cell.

In virology, the term may also be used when referring to mRNA synthesis from an RNA molecule (i.e., RNA replication). For instance, the genome of a negative-sense single-stranded RNA (ssRNA -) virus may be template for a positive-sense single-stranded RNA (ssRNA +). This is because the positive-sense strand contains the information needed to translate the viral proteins for viral replication afterwards. This process is catalyzed by a viral RNA replicase.

A DNA transcription unit encoding for a protein may contain both a *coding sequence*, which will be translated into the protein, and *regulatory sequences*, which direct and regulate the synthesis of that protein. The regulatory sequence before ("upstream" from) the coding sequence is called the five prime untranslated region (5'UTR); the sequence after ("downstream" from) the coding sequence is called the three prime untranslated region (3'UTR).

As opposed to DNA replication, transcription results in an RNA complement that includes the nucleotide uracil (U) in all instances where thymine (T) would have occurred in a DNA complement.

Only one of the two DNA strands serve as a template for transcription. The antisense strand of

DNA is read by RNA polymerase from the 3' end to the 5' end during transcription (3' → 5'). The complementary RNA is created in the opposite direction, in the 5' → 3' direction, matching the sequence of the sense strand with the exception of switching uracil for thymine. This directionality is because RNA polymerase can only add nucleotides to the 3' end of the growing mRNA chain. This use of only the 3' → 5' DNA strand eliminates the need for the Okazaki fragments that are seen in DNA replication. This also removes the need for an RNA primer to initiate RNA synthesis, as is the case in DNA replication.

The non-template (sense) strand of DNA is called the coding strand, because its sequence is the same as the newly created RNA transcript (except for the substitution of uracil for thymine). This is the strand that is used by convention when presenting a DNA sequence.

Transcription has some proofreading mechanisms, but they are fewer and less effective than the controls for copying DNA. As a result, transcription has a lower copying fidelity than DNA replication.

Major Steps

Transcription is divided into initiation, promoter escape, elongation, and termination.

Initiation

Transcription begins with the binding of RNA polymerase, together with one or more general transcription factors, to a specific DNA sequence referred to as a "promoter" to form an RNA polymerase-promoter "closed complex". In the "closed complex" the promoter DNA is still fully double-stranded.

RNA polymerase, assisted by one or more general transcription factors, then unwinds approximately 14 base pairs of DNA to form an RNA polymerase-promoter "open complex". In the "open complex" the promoter DNA is partly unwound and single-stranded. The exposed, single-stranded DNA is referred to as the "transcription bubble."

RNA polymerase, assisted by one or more general transcription factors, then selects a transcription start site in the transcription bubble, binds to an initiating NTP and an extending NTP (or a short RNA primer and an extending NTP) complementary to the transcription start site sequence, and catalyzes bond formation to yield an initial RNA product.

In bacteria, RNA polymerase holoenzyme consists of five subunits: 2 α subunits, 1 β subunit, 1 β' subunit, and 1 ω subunit. In bacteria, there is one general RNA transcription factor known as a sigma factor. RNA polymerase core enzyme binds to the bacterial general transcription (sigma) factor to form RNA polymerase holoenzyme and then binds to a promoter. (RNA polymerase is called a holoenzyme when sigma subunit is attached to the core enzyme which is consist of 2 α subunits, 1 β subunit, 1 β' subunit only).

In archaea and eukaryotes, RNA polymerase contains subunits homologous to each of the five RNA polymerase subunits in bacteria and also contains additional subunits. In archaea and eukaryotes, the functions of the bacterial general transcription factor sigma are performed by multiple general transcription factors that work together.

In archaea, there are three general transcription factors: TBP, TFB, and TFE. In eukaryotes, in RNA polymerase II-dependent transcription, there are six general transcription factors: TFIIA, TFIIB (an ortholog of archaeal TFB), TFIID (a multisubunit factor in which the key subunit, TBP, is an ortholog of archaeal TBP), TFIIE (an ortholog of archaeal TFE), TFIIF, and TFIIH. In archaea and eukaryotes, the RNA polymerase-promoter closed complex is usually referred to as the "preinitiation complex."

Transcription initiation is regulated by additional proteins, known as activators and repressors, and, in some cases, associated coactivators or corepressors, which modulate formation and function of the transcription initiation complex.

Promoter Escape

After the first bond is synthesized, the RNA polymerase must escape the promoter. During this time there is a tendency to release the RNA transcript and produce truncated transcripts. This is called abortive initiation, and is common for both eukaryotes and prokaryotes. Abortive initiation continues to occur until an RNA product of a threshold length of approximately 10 nucleotides is synthesized, at which point promoter escape occurs and a transcription elongation complex is formed.

Mechanistically, promoter escape occurs through DNA scrunching, providing the energy needed to break interactions between RNA polymerase holoenzyme and the promoter.

In bacteria, it was historically thought that the sigma factor is definitely released after promoter clearance occurs. This theory had been known as the *obligate release model* however later data showed that upon and following promoter clearance, the sigma factor is released according to a stochastic model known as the stochastic release model.

In eukaryotes, at an RNA polymerase II-dependent promoter, upon promoter clearance, TFIIH phosphorylates serine 5 on the carboxy terminal domain of RNA polymerase II, leading to the recruitment of capping enzyme (CE). The exact mechanism of how CE induces promoter clearance in eukaryotes is not yet known.

Elongation

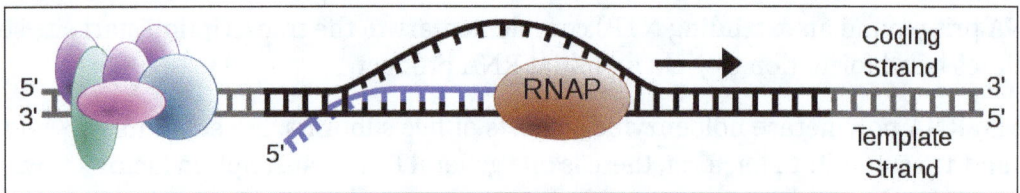

Simple diagram of transcription elongation.

One strand of the DNA, the *template strand* (or noncoding strand), is used as a template for RNA synthesis. As transcription proceeds, RNA polymerase traverses the template strand and uses base pairing complementarity with the DNA template to create an RNA copy (which elongates during the traversal). Although RNA polymerase traverses the template strand from 3' → 5', the coding (non-template) strand and newly formed RNA can also be used as reference points, so transcription can be described as occurring 5' → 3'. This produces an RNA molecule from 5' → 3', an exact copy of the coding strand (except that thymines are replaced with uracils, and the nucleotides are composed of a ribose (5-carbon) sugar where DNA has deoxyribose (one fewer oxygen atom) in its sugar-phosphate backbone).

The mRNA transcription can involve multiple RNA polymerases on a single DNA template and multiple rounds of transcription (amplification of particular mRNA), so many mRNA molecules can be rapidly produced from a single copy of a gene. The characteristic elongation rates in prokaryotes and eukaryotes are about 10-100 nts/sec. In eukaryotes, however, nucleosomes act as major barriers to transcribing polymerases during transcription elongation. In these organisms, the pausing induced by nucleosomes can be regulated by transcription elongation factors such as TFIIS.

Elongation also involves a proofreading mechanism that can replace incorrectly incorporated bases. In eukaryotes, this may correspond with short pauses during transcription that allow appropriate RNA editing factors to bind. These pauses may be intrinsic to the RNA polymerase or due to chromatin structure.

Termination

Bacteria use two different strategies for transcription termination – Rho-independent termination and Rho-dependent termination. In Rho-independent transcription termination, RNA transcription stops when the newly synthesized RNA molecule forms a G-C-rich hairpin loop followed by a run of Us. When the hairpin forms, the mechanical stress breaks the weak rU-dA bonds, now filling the DNA–RNA hybrid. This pulls the poly-U transcript out of the active site of the RNA polymerase, terminating transcription. In the "Rho-dependent" type of termination, a protein factor called "Rho" destabilizes the interaction between the template and the mRNA, thus releasing the newly synthesized mRNA from the elongation complex.

Transcription termination in eukaryotes is less well understood than in bacteria, but involves cleavage of the new transcript followed by template-independent addition of adenines at its new 3' end, in a process called polyadenylation.

Inhibitors

Transcription inhibitors can be used as antibiotics against, for example, pathogenic bacteria (antibacterials) and fungi (antifungals). An example of such an antibacterial is rifampicin, which inhibits bacterial transcription of DNA into mRNA by inhibiting DNA-dependent RNA polymerase by binding its beta-subunit, while 8-hydroxyquinoline is an antifungal transcription inhibitor. The effects of histone methylation may also work to inhibit the action of transcription.

Endogenous Inhibitors

In vertebrates, the majority of gene promoters contain a CpG island with numerous CpG sites. When many of a gene's promoter CpG sites are methylated the gene becomes inhibited (silenced). Colorectal cancers typically have 3 to 6 driver mutations and 33 to 66 hitchhiker or passenger mutations. However, transcriptional inhibition (silencing) may be of more importance than mutation in causing progression to cancer. For example, in colorectal cancers about 600 to 800 genes are transcriptionally inhibited by CpG island methylation. Transcriptional repression in cancer can also occur by other epigenetic mechanisms, such as altered expression of microRNAs. In breast cancer, transcriptional repression of BRCA1 may occur more frequently by over-expressed microRNA-182 than by hypermethylation of the BRCA1 promoter.

Transcription Factors

Active transcription units are clustered in the nucleus, in discrete sites called transcription factories or euchromatin. Such sites can be visualized by allowing engaged polymerases to extend their transcripts in tagged precursors (Br-UTP or Br-U) and immuno-labeling the tagged nascent RNA. Transcription factories can also be localized using fluorescence in situ hybridization or marked by antibodies directed against polymerases. There are ~10,000 factories in the nucleoplasm of a HeLa cell, among which are ~8,000 polymerase II factories and ~2,000 polymerase III factories. Each polymerase II factory contains ~8 polymerases. As most active transcription units are associated with only one polymerase, each factory usually contains ~8 different transcription units. These units might be associated through promoters and enhancers, with loops forming a "cloud" around the factor.

Measuring and Detecting

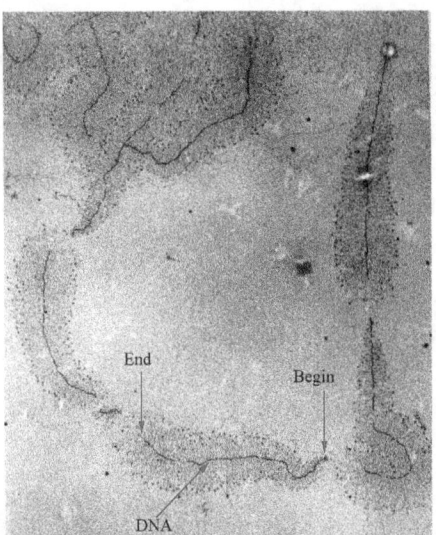

Electron micrograph of transcription of ribosomal RNA. The forming ribosomal RNA strands are visible as branches from the main DNA strand.

Transcription can be measured and detected in a variety of ways:

- G-Less Cassette transcription assay: Measures promoter strength.

- Run-off transcription assay: Identifies transcription start sites (TSS).

- Nuclear run-on assay: Measures the relative abundance of newly formed transcripts.

- RNase protection assay and ChIP-Chip of RNAP: Detect active transcription sites.

- RT-PCR: Measures the absolute abundance of total or nuclear RNA levels, which may however differ from transcription rates.

- DNA microarrays: Measures the relative abundance of the global total or nuclear RNA levels; however, these may differ from transcription rates.

- In situ hybridization: Detects the presence of a transcript.

- MS2 tagging: By incorporating RNA stem loops, such as MS2, into a gene, these become incorporated into newly synthesized RNA. The stem loops can then be detected using a fusion of GFP and the MS2 coat protein, which has a high affinity, sequence-specific interaction with the MS2 stem loops. The recruitment of GFP to the site of transcription is visualized as a single fluorescent spot. This new approach has revealed that transcription occurs in discontinuous bursts, or pulses. With the notable exception of in situ techniques, most other methods provide cell population averages, and are not capable of detecting this fundamental property of genes.

- Northern blot: The traditional method, and until the advent of RNA-Seq, the most quantitative.

- RNA-Seq: Applies next-generation sequencing techniques to sequence whole transcriptomes, which allows the measurement of relative abundance of RNA, as well as the detection of additional variations such as fusion genes, post-transcriptional edits and novel splice sites.

Reverse Transcription

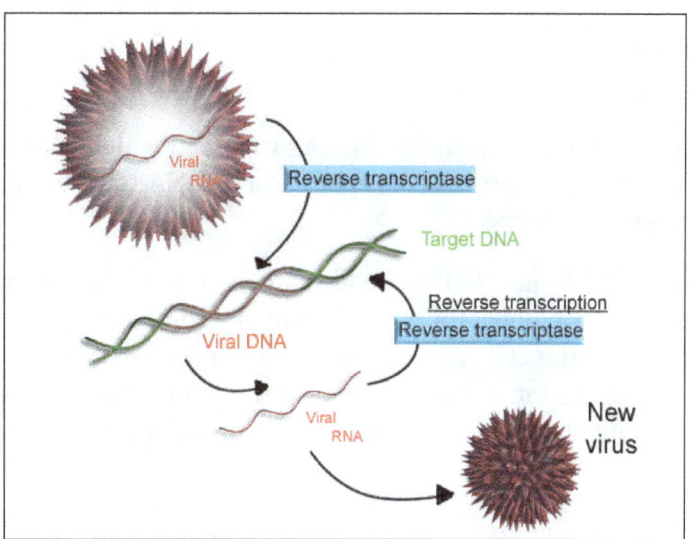

Scheme of reverse transcription.

Some viruses (such as HIV, the cause of AIDS), have the ability to transcribe RNA into DNA. HIV has an RNA genome that is *reverse transcribed* into DNA. The resulting DNA can be merged with the DNA genome of the host cell. The main enzyme responsible for synthesis of DNA from an RNA template is called reverse transcriptase.

In the case of HIV, reverse transcriptase is responsible for synthesizing a complementary DNA strand (cDNA) to the viral RNA genome. The enzyme ribonuclease H then digests the RNA strand, and reverse transcriptase synthesises a complementary strand of DNA to form a double helix DNA structure ("cDNA"). The cDNA is integrated into the host cell's genome by the enzyme integrase, which causes the host cell to generate viral proteins that reassemble into new viral particles. In HIV, subsequent to this, the host cell undergoes programmed cell death, or apoptosis of T cells. However, in other retroviruses, the host cell remains intact as the virus buds out of the cell.

Some eukaryotic cells contain an enzyme with reverse transcription activity called telomerase. Telomerase is a reverse transcriptase that lengthens the ends of linear chromosomes. Telomerase carries an RNA template from which it synthesizes a repeating sequence of DNA, or "junk" DNA. This repeated sequence of DNA is called a telomere and can be thought of as a "cap" for a chromosome. It is important because every time a linear chromosome is duplicated, it is shortened. With this "junk" DNA or "cap" at the ends of chromosomes, the shortening eliminates some of the non-essential, repeated sequence rather than the protein-encoding DNA sequence, that is farther away from the chromosome end.

Telomerase is often activated in cancer cells to enable cancer cells to duplicate their genomes indefinitely without losing important protein-coding DNA sequence. Activation of telomerase could be part of the process that allows cancer cells to become *immortal*. The immortalizing factor of cancer via telomere lengthening due to telomerase has been proven to occur in 90% of all carcinogenic tumors *in vivo* with the remaining 10% using an alternative telomere maintenance route called ALT or Alternative Lengthening of Telomeres.

TRANSLATION

In molecular biology and genetics, translation is the process in which ribosomes in the cytoplasm or ER synthesize proteins after the process of replication of DNA to RNA in the cell's nucleus. The entire process is called gene expression.

In translation, messenger RNA (mRNA) is decoded in the ribosome decoding center to produce a specific polypeptide chain, or polypeptide. The polypeptide later folds into an active protein and performs its functions in the cell. The ribosome facilitates decoding by inducing the binding of complementary tRNA anticodon sequences to rRNA codons. The tRNAs carry specific amino acids that are chained together into a polypeptide as the mRNA passes through and is read by the ribosome.

Translation proceeds in three phases:

1. Initiation: The ribosome assembles around the target mRNA. The first tRNA is attached at the start codon.

2. Elongation: The tRNA transfers an amino acid to the tRNA corresponding to the next codon. The ribosome then moves (*translocates)* to the next mRNA codon to continue the process, creating an amino acid chain.

3. Termination: When a peptidyl tRNA encounters a stop codon, then the ribosome folds the polypeptide into its final structure.

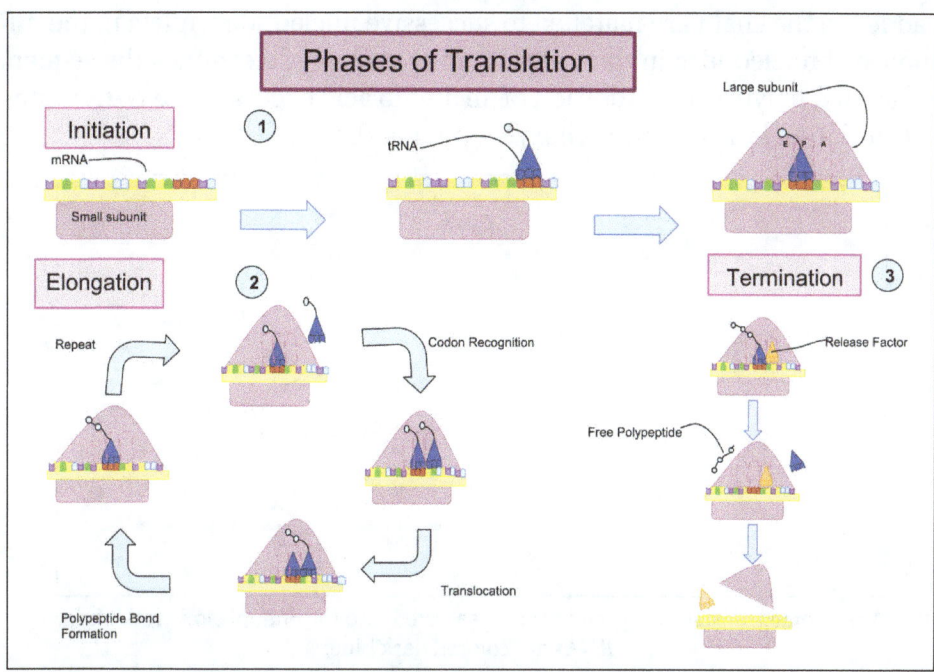

The three phases of translation initiation polymerase binds to the DNA strand and moves along until the small ribosomal subunit binds to the DNA. Elongation is initiated when the large subunit attaches and termination ends the process of elongation.

In prokaryotes (bacteria), translation occurs in the cytosol, where the medium and small subunits of the ribosome bind to the tRNA. In eukaryotes, translation occurs in the cytosol or across the membrane of the endoplasmic reticulum in a process called transposition. In transposition, the entire ribosome/mRNA complex binds to the outer membrane of the rough endoplasmic reticulum (ER) and the new protein is synthesized and released into the ER; the newly created polypeptide can be stored inside the ER for future vesicle transport and secretion outside the cell, or immediately secreted.

Many types of transcribed RNA, such as transfer RNA, ribosomal RNA, and small nuclear RNA, do not undergo translation into proteins.

A number of antibiotics act by inhibiting translation. These include clindamycin, anisomycin, cycloheximide, chloramphenicol, tetracycline, streptomycin, erythromycin, and puromycin. Prokaryotic ribosomes have a different structure from that of eukaryotic ribosomes, and thus antibiotics can specifically target bacterial infections without any harm to a eukaryotic host's cells.

Basic Mechanisms

The basic process of translation is the addition of one amino acid at a time to the end of the polypeptide being formed. This process takes place inside the ribosome. A ribosome is made up of two subunits, a small 40S subunit and a large 60S subunit. These subunits come together before translation of mRNA into a protein to provide a location for translation to be carried out and a polypeptide to be produced. The choice of amino acid type to be added is determined by the genetic code on the mRNA molecule. Each amino acid added is matched to a three nucleotide subsequence of the mRNA. For each such triplet possible, the corresponding amino acid is accepted. The successive amino acids added to the chain are matched to successive nucleotide triplets in the mRNA. In this way, the sequence of nucleotides in the template mRNA chain determines the sequence of amino acids in the generated polypeptide. Addition of an amino acid occurs at the N-terminus of the peptide and thus translation is said to be carboxyl-to-amino directed.

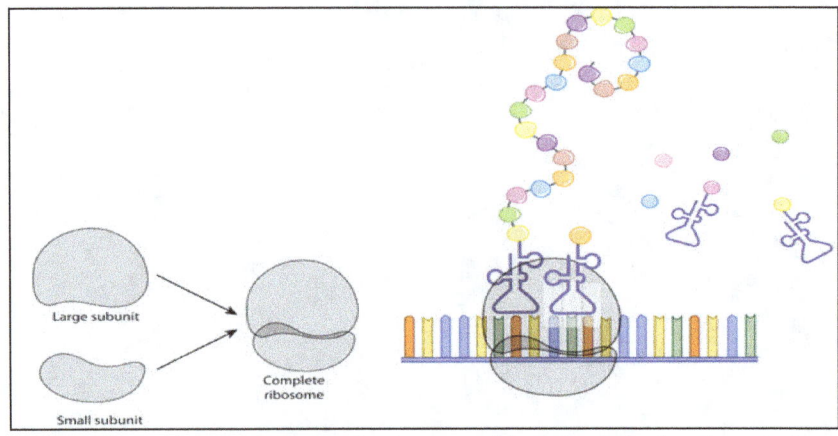

A ribosome translating a protein that is secreted into the endoplasmic reticulum. tRNAs are colored dark blue.

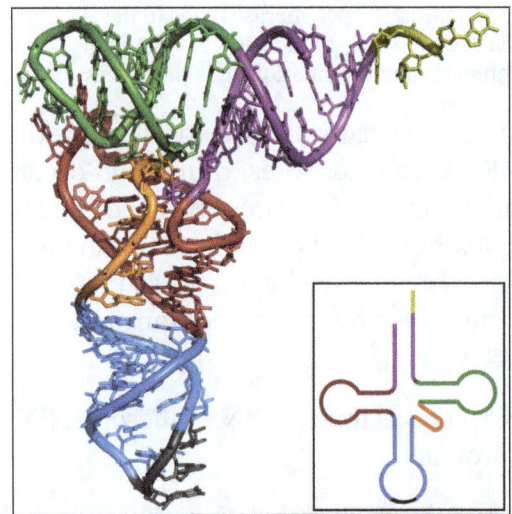

Tertiary structure of tRNA. *CCA tail* in yellow, *Acceptor stem* in purple, *Variable loop* in orange, *D arm* in red, *Anticodon arm* in blue with *Anticodon* in black, *T arm* in green.

The mRNA carries genetic information encoded as a DNA sequence from the chromosomes to the nucleolus. The ribonucleotides are "read" by translational machinery in a sequence of nucleotide triplets called codons. Each of those triplets codes for a specific amino acid.

The ribosome molecules translate this code to a specific sequence of amino acids. The ribosome is a multi-subunit structure containing tRNA and proteins. It is the "factory" where amino acids are assembled into proteins. tRNAs are small noncoding RNA chains (75-90 nucleotides) that transport amino acids to the ribosome. tRNAs have a site for amino acid attachment, and a site called a codon. The codon is an RNA triplet complementary to the mRNA triplet that codes for their cargo amino acid.

Aminoacyl tRNA synthetases (enzymes) catalyze the bonding between specific tRNAs and the amino acids that their codon sequences call for. The product of this reaction is an aminoacyl-tRNA. In prokaryotes, this aminoacyl-tRNA is carried to the ribosome by EF-Tu, where mRNA codons are matched through complementary base pairing to specific tRNA anticodons. Aminoacyl-tRNA synthetases that mispair tRNAs with the wrong amino acids can produce mischarged aminoacyl-tRNAs, which can result in inappropriate amino acids at the respective position in protein. This "mistranslation" of the genetic code naturally occurs at low levels in most organisms, but certain cellular environments cause an increase in permissive mRNA decoding, sometimes to the benefit of the cell.

The ribosome has three sites for tRNA to bind. They are the aminoacyl site (abbreviated A), the peptidyl site (abbreviated P) and the entrance site (abbreviated E). With respect to the mRNA, the three sites are oriented 3' to 5' E-P-A, because ribosomes move toward the 5' end of mRNA. The A-site binds the incoming tRNA with the complementary codon on the mRNA. The P-site holds the tRNA with the growing polypeptide chain. The E-site holds the tRNA without its amino acid, and the tRNA is then released. When an aminoacyl-tRNA initially binds to its corresponding codon on the mRNA, it is in the A site. Then, a peptide bond forms between the amino acid of the tRNA in the A site and the amino acid of the charged tRNA in the P site. The growing polypeptide chain is transferred to the tRNA in the A site. Translocation occurs, moving the tRNA in the P site, now without an amino acid, to the E site; the tRNA that was in the A site, now charged with the polypeptide chain, is moved to the P site. The tRNA in the E site leaves and another aminoacyl-tRNA enters the P site to repeat the process.

After the new amino acid is added to the chain, and after the mRNA is released out of the cytoplasm and into the ribosome's core, the energy provided by the hydrolysis of a GDP bound to the translocase EF-G (in prokaryotes) and eEF-2 (in eukaryotes) moves the ribosome down one codon towards the 3' end. The energy required for translation of proteins is significant. For a protein containing n amino acids, the number of high-energy phosphate bonds required to translate it is $4n+1$. The rate of translation varies; it is significantly higher in prokaryotic cells (up to 17-21 amino acid residues per second) than in eukaryotic cells (up to 6-9 amino acid residues per second).

Even though the ribosomes are usually considered accurate, processive machines, the translation process is subject to errors that can lead either to the synthesis of erroneous proteins or to the premature abandonment of translation. The rate of error in synthesizing proteins has been estimated to be between $1/10^5$ and $1/10^3$ misincorporated amino acids, depending on the experimental conditions. The rate of premature translation abandonment, instead, has been estimated to be of the order of magnitude of 10^{-4} events per translated codon. The correct amino acid is covalently bonded to the correct transfer RNA (tRNA) by amino acyl transferases. The amino acid is joined by its functional group to the 3' OH of the tRNA by an ether bond. When the tRNA has an amino acid linked to it, the tRNA is termed "charged". Initiation involves the small subunit of the ribosome binding to the 5' end of mRNA with the help of initiation factors (IF). In prokaryotes, initiation of protein synthesis involves the recognition of a purine-rich initiation sequence on the tRNA called the Shine-Dalgarno sequence. The Shine-Dalgarno sequence binds to a complementary pyrimidine-rich sequence on the 3' end of the 16S rRNA part of the 30S ribosomal subunit. The binding of these complementary sequences ensures that the 30S ribosomal subunit is bound to the mRNA and is aligned such that the initiation codon is placed in the 30S portion of the P-site. Once the mRNA and 30S subunit are properly bound, an initiation factor brings the initiator tRNA-amino acid complex, f-Met-tRNA, to the 30S P site. The initiation phase is completed once a 50S subunit joins the 30 subunit, forming an active 70S ribosome. Termination of the polypeptide occurs when the A site of the ribosome is occupied by a stop codon (UAA, UAG, or UGA) on the mRNA. mRNA usually can recognize or bind to stop codons. Instead, the stop codon induces the binding of a release factor protein (RF1 & RF2) that prompts the disassembly of the entire ribosome/mRNA complex by the hydrolysis of the polypeptide chain from the peptidyl transferase center of the ribosome. Drugs or special sequence motifs on the mRNA can change the ribosomal structure so that near-cognate tRNAs are bound to the stop codon instead of the release factors. In such cases of 'translational read-through', translation continues until the ribosome encounters the next stop codon.

The process of translation is highly regulated in prokaryotic and eukaryotic organisms. Regulation of translation can impact the global rate of protein synthesis which is closely coupled to the metabolic and proliferative state of a cell. In addition, recent work has revealed that genetic differences and their subsequent expression as mRNAs can also impact translation rate in an RNA-specific manner.

Genetic Code

Whereas other aspects such as the 3D structure, called tertiary structure, of protein can only be predicted using sophisticated algorithms, the amino acid sequence, called primary structure, can be determined solely from the nucleic acid sequence with the aid of a translation table.

This approach may not give the correct amino acid composition of the protein, in particular if unconventional amino acids such as selenocysteine are incorporated into the protein, which is coded for by a conventional stop codon in combination with a downstream hairpin (SElenoCysteine Insertion Sequence, or SECIS).

There are many computer programs capable of translating a DNA/RNA sequence into a protein sequence. Normally this is performed using the Standard Genetic Code, however, few programs can handle all the "special" cases, such as the use of the alternative initiation codons. For instance, the rare alternative start codon CTG codes for Methionine when used as a start codon, and for Leucine in all other positions.

Example: Condensed translation table for the Standard Genetic Code.

```
AAs    = FFLLSSSSYY**CC*WLLLLPPPPHHQQRRRRIIIMTTTTNNKKSSRRVVVVAAAADDEEGGGG

Starts = ---M---------------M---------------M----------------------------

Base1  = TTTTTTTTTTTTTTTTCCCCCCCCCCCCCCCCAAAAUAAAAAAAAAAAAGGGGGGGGGGGGGGGG

Base2  = TTTTCCCCAAAAGGGGTTTTCCCCAAAAGGGGTTTACCCCAAAAGGGGTTTTCCCCAAAAGGGG

Base3  = TCAGTCAGTCAGTCAGTCAGTCAGTCAGTCAGTCAGTCAGTCAGTCAGTCAGTCAGTCAGTCAG
```

The "Starts" row indicate three start codons, UUG, CUG, and the very common UAG. It also indicates the first amino acid residue when interpreted as a start: in this case it is all methionine.

Translation Tables

Even when working with ordinary prokaryotic sequences such as the E.coli genome, it is often desired to be able to use alternative translation tables—namely for transcription of the mitochondrial genes. Currently the following translation tables are defined by the NCBI Taxonomy Group for the translation of the sequences in GenBank:

- The Standard.

- The Vertebrate Mitochondrial Code.

- The Yeast Mitochondrial Code.

- The Mold, Protozoan, and Coelenterate Mitochondrial Code and the Mycoplasma/Spiroplasma Code.

- The Invertebrate Mitochondrial Code.

- The Ciliate, Dasycladacean and Hexamita Nuclear Code.

- The Echinoderm and Flatworm Mitochondrial Code.

- The Euplotid Nuclear Code.

- The Bacterial and Plant Plastid Code.

- The Alternative Yeast Nuclear Code.

- The Ascidian Mitochondrial Code.

- The Alternative Flatworm Mitochondrial Code.

- Blepharisma Nuclear Code.

- Chlorophycean Mitochondrial Code.

- Trematode Mitochondrial Code.

- Scenedesmus obliquus mitochondrial Code.

- Thraustochytrium Mitochondrial Code.

- Pterobranchia Mitochondrial Code.

- Candidate Division SR1 and Gracilibacteria Code.

- Pachysolen tannophilus Nuclear Code.

- Karyorelict Nuclear Code.

- Condylostoma Nuclear Code.

- Mesodinium Nuclear Code.

- Peritrich Nuclear Code.

- Blastocrithidia Nuclear Code.

- Cephalodiscidae Mitochondrial UAA-Tyr Code.

References

- Cobb M (September 2017). "60 years ago, Francis Crick changed the logic of biology". Plos Biology. 15 (9): e2003243. Doi:10.1371/journal.pbio.2003243. PMC 5602739. PMID 28922352

- Geneexpression-regulation, topics, highereducation: projects, le.ac.uk, Retrieved 22 August, 2019

- Ridley R (2001). "What Would Thomas Henry Huxley Have Made of Prion Diseases?". In Baker HF (ed.). Molecular Pathology of the Prions. Methods in Molecular Medicine. Humana Press. Pp. 1–16. ISBN 0-89603-924-2

- "Will the Hayflick limit keep us from living forever?". Howstuffworks. 2009-05-11. Retrieved January 20, 2015

- Recombinant-DNA-technology, science: britannica.com, Retrieved 23 March, 2019

- Moghal A, Mohler K, Ibba M (November 2014). "Mistranslation of the genetic code". FEBS Letters. 588 (23): 4305–10. doi:10.1016/j.febslet.2014.08.035

Molecular Biology: Techniques and Processes

Many important techniques are used in molecular biology to understand various biological processes. Some of these are DNA sequencing, molecular cloning, gel electrophoresis, polymerase chain reaction, DNA microarray, transfection and ligation. The topics elaborated in this chapter will help in gaining a better perspective about these techniques used in molecular biology.

GEL ELECTROPHORESIS

Gel electrophoresis is a technique used to separate DNA fragments (or other macromolecules, such as RNA and proteins) based on their size and charge. Electrophoresis involves running a current through a gel containing the molecules of interest. Based on their size and charge, the molecules will travel through the gel in different directions or at different speeds, allowing them to be separated from one another.

All DNA molecules have the same amount of charge per mass. Because of this, gel electrophoresis of DNA fragments separates them based on size only. Using electrophoresis, we can see how many different DNA fragments are present in a sample and how large they are relative to one another. We can also determine the absolute size of a piece of DNA by examining it next to a standard "yardstick" made up of DNA fragments of known sizes.

Gel

At one end, the gel has pocket-like indentations called wells, which are where the DNA samples will be placed:

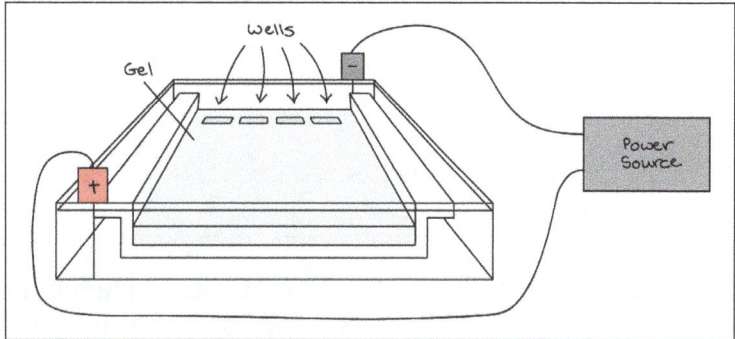

As the name suggests, gel electrophoresis involves a gel: a slab of Jello-like material. Gels for DNA separation are often made out of a polysaccharide called agarose, which comes as dry, powdered flakes. When the agarose is heated in a buffer (water with some salts in it) and allowed to cool, it will form a solid, slightly squishy gel. At the molecular level, the gel is a matrix of agarose molecules that are held together by hydrogen bonds and form tiny pores.

Before the DNA samples are added, the gel must be placed in a gel box. One end of the box is hooked to a positive electrode, while the other end is hooked to a negative electrode. The main body of the box, where the gel is placed, is filled with a salt-containing buffer solution that can conduct current. Although you may not be able to see in the image above, the buffer fills the gel box to a level where it just barely covers the gel.

The end of the gel with the wells is positioned towards the negative electrode. The end without wells (towards which the DNA fragments will migrate) is positioned towards the positive electrode.

Movement of DNA Fragments through the Gel

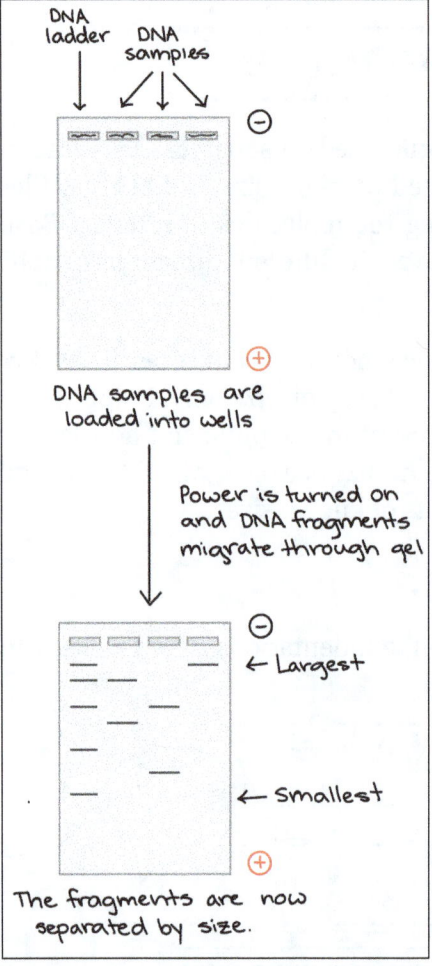

Once the gel is in the box, each of the DNA samples we want to examine (for instance, each PCR reaction or each restriction-digested plasmid) is carefully transferred into one of the wells. One well is reserved for a DNA ladder, a standard reference that contains DNA fragments of known lengths. Commercial DNA ladders come in different size ranges, so we would want to pick one with good "coverage" of the size range of our expected fragments.

Next, the power to the gel box is turned on, and current begins to flow through the gel. The DNA molecules have a negative charge because of the phosphate groups in their sugar-phosphate backbone, so they start moving through the matrix of the gel towards the positive pole. When the power is turned on and current is passing through the gel, the gel is said to be running.

As the gel runs, shorter pieces of DNA will travel through the pores of the gel matrix faster than longer ones. After the gel has run for awhile, the shortest pieces of DNA will be close to the positive end of the gel, while the longest pieces of DNA will remain near the wells. Very short pieces of DNA may have run right off the end of the gel if we left it on for too long.

Visualizing the DNA Fragments

Once the fragments have been separated, we can examine the gel and see what sizes of bands are found on it. When a gel is stained with a DNA-binding dye and placed under UV light, the DNA fragments will glow, allowing us to see the DNA present at different locations along the length of the gel.

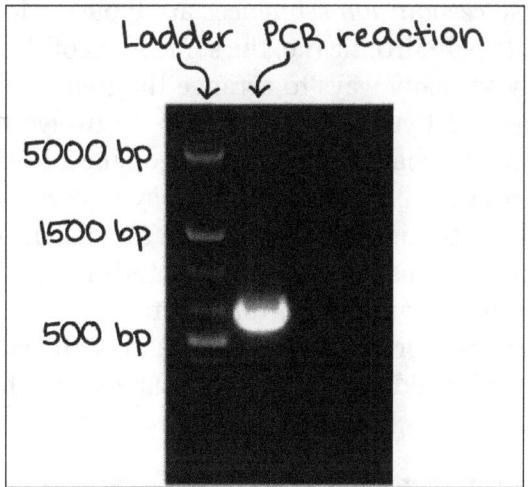

The bp next to each number in the ladder indicates how many base
pairs long the DNA fragment is.

A well-defined "line" of DNA on a gel is called a band. Each band contains a large number of DNA fragments of the same size that have all traveled as a group to the same position. A single DNA fragment (or even a small group of DNA fragments) would not be visible by itself on a gel.

By comparing the bands in a sample to the DNA ladder, we can determine their approximate sizes. For instance, the bright band on the gel above is roughly 700 base pairs (bp) in size.

RESTRICTION DIGEST

A restriction digest is a procedure used in molecular biology to prepare DNA for analysis or other processing. It is sometimes termed *DNA fragmentation* (this term is used for other procedures as well). Hartl and Jones describe it this way:

> "This enzymatic technique can be used for cleaving DNA molecules at specific sites,

ensuring that all DNA fragments that contain a particular sequence at a particular location have the same size; furthermore, each fragment that contains the desired sequence has the sequence located at exactly the same position within the fragment. The cleavage method makes use of an important class of DNA-cleaving enzymes isolated primarily from bacteria. These enzymes are called restriction endonucleases or restriction enzymes, and they are able to cleave DNA molecules at the positions at which particular short sequences of bases are present."

The resulting digested DNA is very often selectively amplified using PCR, making it more suitable for analytical techniques such as agarose gel electrophoresis, and chromatography. It is used in genetic fingerprinting, plasmid subcloning, and RFLP analysis.

Restriction Site

A given restriction enzyme cuts DNA segments within a specific nucleotide sequence, at what is called a restriction site. These *recognition sequences* are typically four, six, eight, ten, or twelve nucleotides long and generally palindromic (i.e. the same nucleotide sequence in the 5' - 3' direction). Because there are only so many ways to arrange the four nucleotides that compose DNA (Adenine, Thymine, Guanine and Cytosine) into a four- to twelve-nucleotide sequence, recognition sequences tend to occur by chance in any long sequence. Restriction enzymes specific to hundreds of distinct sequences have been identified and synthesized for sale to laboratories, and as a result, several potential "restriction sites" appear in almost any gene or locus of interest on any chromosome. Furthermore, almost all artificial plasmids include a (often entirely synthetic) polylinker (also called "multiple cloning site") that contains dozens of restriction enzyme recognition sequences within a very short segment of DNA. This allows the insertion of almost any specific fragment of DNA into plasmid vectors, which can be efficiently "cloned" by insertion into replicating bacterial cells.

After restriction digest, DNA can then be analysed using agarose gel electrophoresis. In gel electrophoresis, a sample of DNA is first "loaded" onto a slab of agarose gel (literally pipetted into small wells at one end of the slab). The gel is then subjected to an electric field, which draws the negatively charged DNA across it. The molecules travel at different rates (and therefore end up at different distances) depending on their net charge (more highly charged particles travel further), and size (smaller particles travel further). Since none of the four nucleotide bases carry any charge, net charge becomes insignificant and size is the main factor affecting rate of diffusion through the gel. Net charge in DNA is produced by the sugar-phosphate backbone. This is in contrast to proteins, in which there is no "backbone", and net charge is generated by different combinations and numbers of charged amino acids.

Possible Uses

Restriction digest is most commonly used as part of the process of the molecular cloning of DNA fragment into a vector (such as a cloning vector or an expression vector). The vector typically contains a multiple cloning site where many restriction site may be found, and a foreign piece of DNA may be inserted into the vector by first cutting the restriction sites in the vector as well the DNA fragment, followed by ligation of the DNA fragment into the vector.

Restriction digests are also necessary for performing any of the following analytical techniques:

- RFLP – Restriction fragment length polymorphism.

- AFLP – Amplified fragment length polymorphism.

- STRP – Short tandem repeat polymorphism.

Various Restriction Enzymes

There are numerous types of restriction enzymes, each of which will cut DNA differently. Most commonly used restriction enzymes are Type II restriction endonuclease. There are some that cut a three base pair sequence while others can cut four, six, and even eight. Each enzyme has distinct properties that determine how efficiently it can cut and under what conditions. Most manufacturers that produce such enzymes will often provide a specific buffer solution that contains the unique mix of cations and other components that aid the enzyme in cutting as efficiently as possible. Different restriction enzymes may also have different optimal temperatures under which they function.

Note that for efficient digest of DNA, the restriction site should not be located at the very end of a DNA fragment. The restriction enzymes may require a minimum number of base pairs between the restriction site and the end of the DNA for the enzyme to work efficiently. This number may vary between enzymes, but for most commonly used restriction enzymes around 6-10 base pair is sufficient.

POLYMERASE CHAIN REACTION

Polymerase chain reaction (PCR) is a common laboratory technique used to make many copies (millions or billions) of a particular region of DNA. This DNA region can be anything the experimenter is interested in. For example, it might be a gene whose function a researcher wants to understand, or a genetic marker used by forensic scientists to match crime scene DNA with suspects.

Typically, the goal of PCR is to make enough of the target DNA region that it can be analyzed or used in some other way. For instance, DNA amplified by PCR may be sent for sequencing, visualized by gel electrophoresis, or cloned into a plasmid for further experiments.

PCR is used in many areas of biology and medicine, including molecular biology research, medical diagnostics, and even some branches of ecology.

Taq Polymerase

Like DNA replication in an organism, PCR requires a DNA polymerase enzyme that makes new strands of DNA, using existing strands as templates. The DNA polymerase typically used in PCR is called Taq polymerase, after the heat-tolerant bacterium from which it was isolated (Thermus aquaticus).

T. aquaticus lives in hot springs and hydrothermal vents. Its DNA polymerase is very heat-stable and is most active around 70 °C (a temperature at which a human or E. coli DNA polymerase would be nonfunctional). This heat-stability makes Taq polymerase ideal for PCR. As we'll see, high temperature is used repeatedly in PCR to denature the template DNA, or separate its strands.

PCR Primers

Like other DNA polymerases, Taq polymerase can only make DNA if it's given a primer, a short sequence of nucleotides that provides a starting point for DNA synthesis. In a PCR reaction, the experimenter determines the region of DNA that will be copied, or amplified, by the primers she or he chooses.

PCR primers are short pieces of single-stranded DNA, usually around 20 nucleotides in length. Two primers are used in each PCR reaction, and they are designed so that they flank the target region (region that should be copied). That is, they are given sequences that will make them bind to opposite strands of the template DNA, just at the edges of the region to be copied. The primers bind to the template by complementary base pairing.

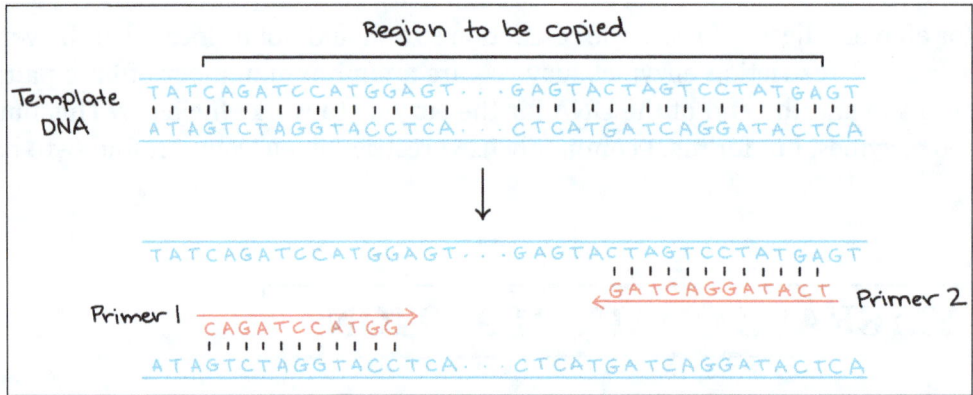

When the primers are bound to the template, they can be extended by the polymerase, and the region that lies between them will get copied.

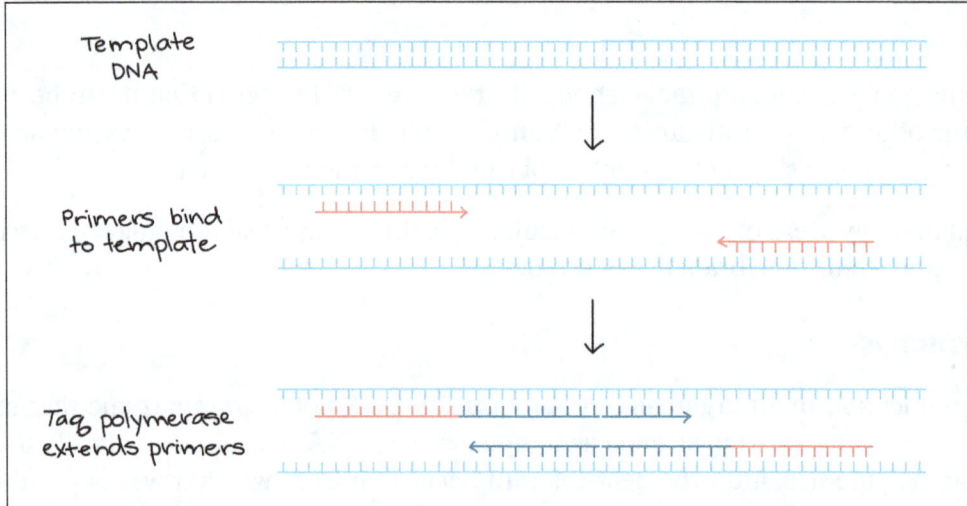

Steps of PCR

The key ingredients of a PCR reaction are Taq polymerase, primers, template DNA, and nucleotides (DNA building blocks). The ingredients are assembled in a tube, along with cofactors needed by the enzyme, and are put through repeated cycles of heating and cooling that allow DNA to be synthesized.

The basic steps are:

1. Denaturation (96 °C): Heat the reaction strongly to separate, or denature, the DNA strands. This provides single-stranded template for the next step.

2. Annealing (55 - 65 °C): Cool the reaction so the primers can bind to their complementary sequences on the single-stranded template DNA.

3. Extension (72 °C): Raise the reaction temperatures so Taq polymerase extends the primers, synthesizing new strands of DNA.

This cycle repeats 25 - 35 times in a typical PCR reaction, which generally takes 2 - 4 hours, depending on the length of the DNA region being copied. If the reaction is efficient (works well), the target region can go from just one or a few copies to billions.

That's because it's not just the original DNA that's used as a template each time. Instead, the new DNA that's made in one round can serve as a template in the next round of DNA synthesis. There are many copies of the primers and many molecules of Taq polymerase floating around in the reaction, so the number of DNA molecules can roughly double in each round of cycling. This pattern of exponential growth is shown in the image below.

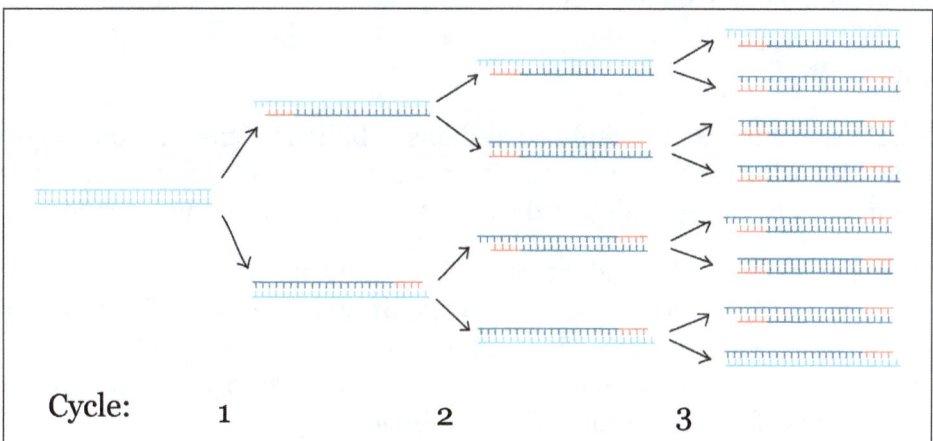

Cycle: 1   2   3

Using Gel Electrophoresis to Visualize the Results of PCR

The results of a PCR reaction are usually visualized (made visible) using gel electrophoresis. Gel electrophoresis is a technique in which fragments of DNA are pulled through a gel matrix by an electric current, and it separates DNA fragments according to size. A standard, or DNA ladder, is typically included so that the size of the fragments in the PCR sample can be determined.

DNA fragments of the same length form a "band" on the gel, which can be seen by eye if the gel is stained with a DNA-binding dye. For example, a PCR reaction producing a 400 base pair (bp) fragment would look like this on a gel:

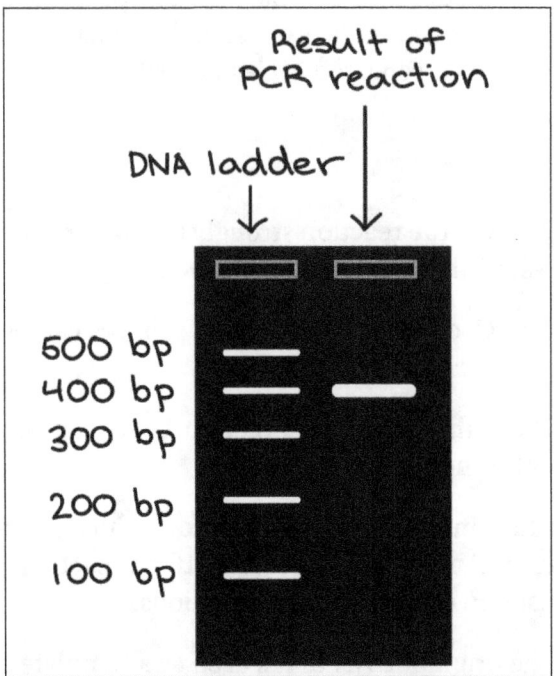

A DNA band contains many, many copies of the target DNA region, not just one or a few copies. Because DNA is microscopic, lots of copies of it must be present before we can see it by eye. This is a big part of why PCR is an important tool: it produces enough copies of a DNA sequence that we can see or manipulate that region of DNA.

Applications of PCR

Using PCR, a DNA sequence can be amplified millions or billions of times, producing enough DNA copies to be analyzed using other techniques. For instance, the DNA may be visualized by gel electrophoresis, sent for sequencing, or digested with restriction enzymes and cloned into a plasmid.

PCR is used in many research labs, and it also has practical applications in forensics, genetic testing, and diagnostics. For instance, PCR is used to amplify genes associated with genetic disorders from the DNA of patients (or from fetal DNA, in the case of prenatal testing). PCR can also be used to test for a bacterium or DNA virus in a patient's body: if the pathogen is present, it may be possible to amplify regions of its DNA from a blood or tissue sample.

LIGATION

In molecular biology, ligation is the joining of two nucleic acid fragments through the action of an enzyme. It is an essential laboratory procedure in the molecular cloning of DNA whereby DNA fragments are joined together to create recombinant DNA molecules, such as when a foreign DNA fragment is inserted into a plasmid. The ends of DNA fragments are joined together by the formation of phosphodiester bonds between the 3'-hydroxyl of one DNA terminus with the 5'-phosphoryl of another. RNA may also be ligated similarly. A co-factor is generally involved in the reaction, and this is usually ATP or NAD^+.

Ligation in the laboratory is normally performed using T4 DNA ligase, however, procedures for ligation without the use of standard DNA ligase are also popular.

Ligation Reaction

The mechanism of the ligation reaction was first elucidated in the laboratory of I. Robert Lehman. Two fragments of DNA may be joined together by DNA ligase which catalyzes the formation of a phosphodiester bond between the 3'-OH at one end of a strand of DNA and the 5'-phosphate group of another. In animals and bacteriophage, ATP is used as the energy source for the ligation, while In bacteria, NAD^+ is used.

The DNA ligase first reacts with ATP or NAD^+, forming a ligase-AMP intermediate with the AMP linked to the ε-amino group of lysine in the active site of the ligase via a phosphoamide bond. This adenylyl group is then transferred to the phosphate group at the 5' end of a DNA chain, forming a DNA-adenylate complex. Finally, a phosphodiester bond between the two DNA ends is formed via the nucleophilic attack of the 3'-hydroxyl at the end of a DNA strand on the activated 5'-phosphoryl group of another.

A nick in the DNA (i.e. a break in one strand of a double-stranded DNA) can be repaired very efficiently by the ligase. However, a complicating feature of ligation presents itself when ligating two separate DNA ends as the two ends need to come together before the ligation reaction can proceed. In the ligation of DNA with sticky or cohesive ends, the protruding strands of DNA may be annealed together already, therefore it is a relatively efficient process as it is equivalent to repairing two nicks in the DNA. However, in the ligation of blunt-ends, which lack protruding ends for the DNA to anneal together, the process is dependent on random collision for the ends to align together before they can be ligated, and is consequently a much less efficient process. The DNA ligase from *E. coli* cannot ligate blunt-ended DNA except under conditions of molecular crowding, and it is therefore not normally used for ligation in the laboratory. Instead the DNA ligase from phage T4 is used as it can ligate blunt-ended DNA as well as single-stranded DNA.

Factors Affecting Ligation

Factors that affect an enzyme-mediated chemical reaction would naturally affect a ligation reaction, such as the concentration of enzyme and the reactants, as well as the temperature of reaction and the length of time of incubation. Ligation is complicated by the fact that the desired ligation products for most ligation reactions should be between two different DNA molecules and the reaction involves both inter- and intra-molecular reactions, and that an additional annealing step is necessary for efficient ligation.

The three steps to form a new phosphodiester bond during ligation are: enzyme adenylylation, adenylyl transfer to DNA, and nick sealing. Mg(2+) is a cofactor for catalysis, therefore at high concentration of Mg(2+) the ligation efficiency is high. If the concentration of Mg(2+) is limited, the nick- sealing is the rate- limiting reaction of the process, and adenylylated DNA intermediate stays in the solution. Such adenylylation of the enzyme restrains the rebinding to the adenylylated DNA intermediate comparison of an Achilles' heel of LIG1, and represents a risk if they are not fixed.

DNA Concentration

The concentration of DNA can affect the rate of ligation, and whether the ligation is an inter-molecular or intra-molecular reaction. Ligation involves joining up the ends of a DNA with other ends, however, each DNA fragment has two ends, and if the ends are compatible, a DNA molecule can circularize by joining its own ends. At high DNA concentration, there is a greater chance of one end of a DNA molecule meeting the end of another DNA, thereby forming intermolecular ligation. At a lower DNA concentration, the chance that one end of a DNA molecule would meet the other end of the same molecule increases, therefore intramolecular reaction that circularizes the DNA is more likely. The transformation efficiency of linear DNA is also much lower than circular DNA, and for the DNA to circularize, the DNA concentration should not be too high. As a general rule, the total DNA concentration should be less than 10 µg/ml.

The relative concentration of the DNA fragments, their length, as well as buffer conditions are also factors that can affect whether intermolecular or intramolecular reactions are favored.

The concentration of DNA can be artificially increased by adding condensing agents such as cobalt hexamine and biogenic polyamines such as spermidine, or by using crowding agents such as polyethylene glycol (PEG) which also increase the effective concentration of enzymes. Note however that additives such as cobalt hexamine can produce exclusively intermolecular reaction, resulting in linear concatemers rather than the circular DNA more suitable for transformation of plasmid DNA, and is therefore undesirable for plasmid ligation. If it is necessary to use additives in plasmid ligation, the use of PEG is preferable as it can promote intramolecular as well as intermolecular ligation.

Ligase Concentration

The higher the ligase concentration, the faster the rate of ligation. Blunt-end ligation is much less efficient than sticky end ligation, so a higher concentration of ligase is used in blunt-end ligations. High DNA ligase concentration may be used in conjunction with PEG for a faster ligation, and they are the components often found in commercial kits designed for rapid ligation.

Temperature

Two issues are involved when considering the temperature of a ligation reaction. First, the optimum temperature for DNA ligase activity which is 37 °C, and second, the melting temperature (T_m) of the DNA ends to be ligated. The melting temperature is dependent on length and base composition of the DNA overhang—the greater the number of G and C, the higher the T_m since there are three hydrogen bonds formed between G-C base pair compared to two for A-T base pair—with some contribution from the stacking of the bases between fragments. For the ligation reaction to proceed efficiently, the ends should be stably annealed, and in ligation experiments, the T_m of the DNA ends is generally much lower than 37 °C. The optimal temperature for ligating cohesive ends is therefore a compromise between the best temperature for DNA ligase activity and the T_m where the ends can associate. However, different restriction enzymes generates different ends, and the base composition of the ends produced by these enzymes may also differ, the melting temperature and therefore the optimal temperature can vary widely depending on the restriction enzymes used, and the optimum temperature for ligation may be between 4-15 °C depending on the ends. Ligations also often involve ligating ends generated from different restriction enzymes in the same reaction mixture, therefore it may not be practical to select optimal temperature for a particular ligation reaction and most protocols simply choose 12-16 °C, room temperature, or 4 °C.

Buffer Composition

The ionic strength of the buffer used can affect the ligation. The kinds of cations presence can also influence the ligation reaction, for example, excess amount of Na^+ can cause the DNA to become more rigid and increase the likelihood of intermolecular ligation. At high concentration of monovalent cation (>200 mM) ligation can also be almost completely inhibited. The standard buffer used for ligation is designed to minimize ionic effects.

Sticky-end Ligation

Restriction enzymes can generate a wide variety of ends in the DNA they digest, but in cloning experiments most commonly-used restriction enzymes generate a 4-base single-stranded overhang called the sticky or cohesive end (exceptions include *Nde*I which generates a 2-base overhang, and those that generate blunt ends). These sticky ends can anneal to other compatible ends and become ligated in a sticky-end (or cohesive end) ligation. *Eco*RI for example generates an AATT end, and since A and T have lower melting temperature than C and G, its melting temperature T_m is low at around 6 °C. For most restriction enzymes, the overhangs generated have a T_m that is around 15 °C. For practical purposes, sticky end ligations are performed at 12-16 °C, or at room temperature, or alternatively at 4 °C for a longer period.

For the insertion of a DNA fragment into a plasmid vector, it is preferable to use two different restriction enzymes to digest the DNA so that different ends are generated. The two different ends can prevent the religation of the vector without any insert, and it also allows the fragment to be inserted in a directional manner.

When it is not possible to use two different sites, then the vector DNA may need to be dephosphorylated to avoid a high background of recircularized vector DNA with no insert. Without a phosphate group at the ends the vector cannot ligate to itself, but can be ligated to an insert with a phosphate group. Dephosphorylation is commonly done using calf-intestinal alkaline phosphatase (CIAP) which removes the phosphate group from the 5' end of digested DNA, but note that CIAP is not easy to inactivate and can interfere with ligation without an additional step to remove the CIAP, thereby resulting in failure of ligation. CIAP should not be used in excessive amount and should only be used when necessary. Shrimp alkaline phosphatase (SAP) or Antarctic phosphatase (AP) are suitable alternative as they can be easily inactivated.

Blunt-end Ligation

Blunt end ligation does not involve base-pairing of the protruding ends, so any blunt end may be ligated to another blunt end. Blunt ends may be generated by restriction enzymes such as *Sma*I and *Eco*RV. A major advantage of blunt-end cloning is that the desired insert does not require any restriction sites in its sequence as blunt-ends are usually generated in a PCR, and the PCR generated blunt-ended DNA fragment may then be ligated into a blunt-ended vector generated from restriction digest.

Blunt-end ligation, however, is much less efficient than sticky end ligation, typically the reaction is 100× slower than sticky-end ligation. Since blunt-end does not have protruding ends, the ligation reaction depends on random collisions between the blunt-ends and is consequently much less efficient. To compensate for the lower efficiency, the concentration of ligase used is higher than sticky end ligation (10× or more). The concentration of DNA used in blunt-end ligation is also higher to increase the likelihood of collisions between ends, and longer incubation time may also be used for blunt-end ligations.

If both ends needed to be ligated into a vector are blunt-ended, then the vector needs to be dephosphorylated to minimize self-ligation. This may be done using CIAP, but caution in its use is necessary as noted previously. Since the vector has been dephosphorylated, and ligation requires the presence of a 5'-phosphate, the insert must be phosphorylated. Blunt-ended PCR product normally lacks a 5'-phosphate, therefore it needs to be phosphorylated by treatment with T4 polynucleotide kinase. Blunt-end ligation is also reversibly inhibited by high concentration of ATP.

PCR usually generates blunt-ended PCR products, but note that PCR using *Taq* polymerase can add an extra adenine (A) to the 3' end of the PCR product. This property may be exploited in TA cloning where the ends of the PCR product can anneal to the T end of a vector. TA ligation is therefore a form of sticky end ligation. Blunt-ended vectors may be turned into vector for TA ligation with dideoxythymidine triphosphate (ddTTP) using terminal transferase.

Guidelines

For the cloning of an insert into a circular plasmid:

- The total DNA concentration used should be less than 10 µg/ml as the plasmid needs to recircularize.

- The molar ratio of insert to vector is usually used at around 3:1. Very high ratio may produce multiple inserts. The ratio may be adjusted depending on the size of the insert, and other ratios may be used, such as 1:1.

Trouble-shooting

Sometimes ligation fail to produce the desired ligated products, and some of the possible reasons may be:

- Damaged DNA – Over-exposure to UV radiation during preparation of DNA for ligation can damage the DNA and significantly reduce transformation efficiency. A higher-wavelength UV radiation (365 nm) which cause less damage to DNA should be used if it is necessary work for work on the DNA on a UV transilluminator for an extended period of time. Addition of cytidine or guanosine to the electrophoresis buffer at 1 mM concentration however may protect the DNA from damage.

- Incorrect usage of CIAP or its inefficient inactivation or removal.

- Excessive amount of DNA used.

- Incomplete DNA digest – The vector DNA that is incompletely digested will give rise to a high background, and this may be checked by doing a ligation without insert as a control. Insert that is not completely digested will also not ligate properly and circularize. When digesting a PCR product, make sure that sufficient extra bases have been added to the 5'-ends of the oligonucleotides used for PCR as many restriction enzymes require a minimum number of extra basepairs for efficient digest. The information on the minimum basepair required is available from restriction enzyme suppliers such as in the catalog of New England Biolabs.

- Incomplete ligation – Blunt-ends DNA (e.g. SmaI) and some sticky-ends DNA (e.g. NdeI) that have low-melting temperature require more ligase and longer incubation time.

- Protein expressed from ligated gene insert is toxic to cells.

- Homologous sequence in insert to sequence in plasmid DNA resulting in deletion.

Other Methods of DNA Ligation

A number of commercially available DNA cloning kits use other methods of ligation that do not require the use of the usual DNA ligases. These methods allow cloning to be done much more rapidly, as well as allowing for simpler transfer of cloned DNA insert to different vectors. These methods however require the use of specially designed vectors and components, and may lack flexibility.

Topoisomerase-mediated Ligation

Topoisomerase can be used instead of ligase for ligation, and the cloning may be done more rapidly without the need for restriction digest of the vector or insert. In this TOPO cloning method a linearized vector is activated by attaching topoisomerase I to its ends, and this "TOPO-activated" vector may then accept a PCR product by ligating to both of the 5' ends of the PCR product, the topoisomerase is released and a circular vector is formed in the process.

Homologous Recombination

Another method of cloning without the use of ligase is by DNA recombination, for example as used in the Gateway cloning system. The gene, once cloned into the cloning vector (called entry clone in this method), may be conveniently introduced into a variety of expression vectors by recombination.

BLOT

A blot, in molecular biology and genetics, is a method of transferring proteins, DNA or RNA, onto a carrier (for example, a nitrocellulose, polyvinylidene fluoride (PVDF) or nylon membrane). In many instances, this is done after a gel electrophoresis, transferring the molecules from the gel onto the blotting membrane, and other times adding the samples directly onto the membrane. After the blotting, the transferred proteins, DNA or RNA are then visualized by colorant staining (for example, silver staining of proteins), autoradiographic visualization of radioactive labelled molecules (performed before the blot), or specific labelling of some proteins or nucleic acids. The latter is done with antibodies or hybridization probes that bind only to some molecules of the blot and have an enzyme joined to them. After proper washing, this enzymatic activity (and so, the molecules we search in the blot) is visualized by incubation with proper reactive, rendering either a colored deposit on the blot or a chemiluminiscent reaction which is registered by photographic film.

Southern Blot

A Southern blot is a method routinely used in molecular biology for detection of a specific DNA sequence in DNA samples. Southern blotting combines transfer of electrophoresis-separated DNA fragments to a filter membrane and subsequent fragment detection by probe hybridization.

Western Blot

A Western blot is used for the detection of specific proteins in complex samples. Proteins are first separated by size using electrophoresis before being transferred to an appropriate blotting matrix (usually PVDF or nitrocellulose) and subsequent detection with antibodies.

Far-Western Blot

Similar to a Western blot, the Far-Western blot uses protein–protein interactions to detect the presence of a specific protein immobilized on a blotting matrix. Antibodies are then used to detect the presence of the protein–protein complex, making the Far-Western blot a specific case of the Western blot. .

Southwestern Blot

A Southwestern blot is based on Southern blotting and is used to identify and characterize DNA-binding proteins by their ability to bind to specific oligonucleotide probes. The proteins are separated by gel electrophoresis and are subsequently transferred to nitrocellulose membranes similar to other types of blotting.

Eastern Blot

The Eastern blot is used for the detection of specific posttranslational modifications of proteins. Proteins are separated by gel electrophoresis before being transferred to a blotting matrix whereupon posttranslational modifications are detected by specific substrates (cholera toxin, concanavalin, phosphomolybdate, etc.) or antibodies.

Far-Eastern Blot

The Far-Eastern blot is for the detection of lipid-linked oligosaccharides. High performance thin layer chromatography is first used to separate the lipids by physical and chemical characteristics, then transferred to a blotting matrix before the oligosaccharides are detected by a specific binding protein (i.e. antibodies or lectins).

Northern Blot

The Northern blot is for the detection of specific RNA sequences in complex samples. Northern blotting first separates samples by size via gel electrophoresis before they are transferred to a blotting matrix and detected with labeled RNA probes.

Reverse Northern Blot

The Reverse Northern blot differs from both Northern and Southern blotting in that DNA is first immobilized on a blotting matrix and specific sequences are detected with labeled RNA probes.

Dot Blot

A Dot blot is a special case of any of the above blots where the analyte is added directly to the blotting matrix (and appears as a "dot") as opposed to separating the sample by electrophoresis prior to blotting.

List of Blots

- Northern blot for RNA.

- Reverse Northern blot for RNA.

- Western blot for proteins.

- Far-Western blot for protein–protein interactions.

- Eastern blotting for posttranslational modification.

- Far-Eastern blot for glycolipids.

- Dot blot.

- Southern blot for DNA.

DNA SEQUENCING

DNA sequencing is the process of determining the nucleic acid sequence – the order of nucleotides in DNA. It includes any method or technology that is used to determine the order of the four bases: adenine, guanine, cytosine, and thymine. The advent of rapid DNA sequencing methods has greatly accelerated biological and medical research and discovery.

Knowledge of DNA sequences has become indispensable for basic biological research, and in numerous applied fields such as medical diagnosis, biotechnology, forensic biology, virology and biological systematics. Comparing healthy and mutated DNA sequences can diagnose different diseases including various cancers, characterize antibody repertoire, and can be used to guide patient treatment. Having a quick way to sequence DNA allows for faster and more individualized medical care to be administered, and for more organisms to be identified and cataloged.

The rapid speed of sequencing attained with modern DNA sequencing technology has been instrumental in the sequencing of complete DNA sequences, or genomes, of numerous types and species of life, including the human genome and other complete DNA sequences of many animal, plant, and microbial species.

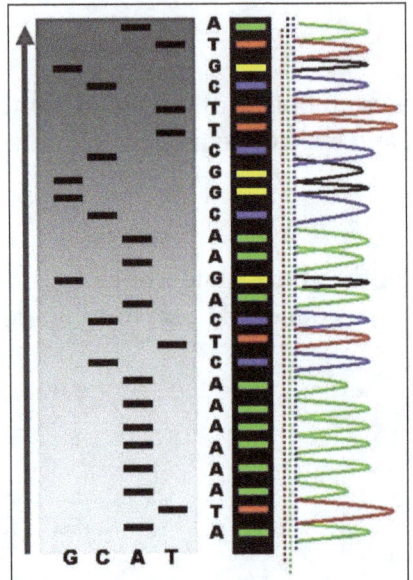

An example of the results of automated chain-termination DNA sequencing.

The first DNA sequences were obtained in the early 1970s by academic researchers using laborious methods based on two-dimensional chromatography. Following the development of fluorescence-based sequencing methods with a DNA sequencer, DNA sequencing has become easier and orders of magnitude faster.

Four Canonical Bases

The canonical structure of DNA has four bases: thymine (T), adenine (A), cytosine (C), and guanine (G). DNA sequencing is the determination of the physical order of these bases in a molecule of DNA. However, there are many other bases that may be present in a molecule. In some viruses (specifically, bacteriophage), cytosine may be replaced by hydroxy methyl or hydroxy methyl glucose cytosine. In mammalian DNA, variant bases with methyl groups or phosphosulfate may be found. Depending on the sequencing technique, a particular modification, e.g., the 5mC (5 methyl cytosine) common in humans, may or may not be detected.

Basic Methods

Maxam-Gilbert Sequencing

Allan Maxam and Walter Gilbert published a DNA sequencing method in 1977 based on chemical modification of DNA and subsequent cleavage at specific bases. Also known as chemical sequencing, this method allowed purified samples of double-stranded DNA to be used without further cloning. This method's use of radioactive labeling and its technical complexity discouraged extensive use after refinements in the Sanger methods had been made.

Maxam-Gilbert sequencing requires radioactive labeling at one 5' end of the DNA and purification of the DNA fragment to be sequenced. Chemical treatment then generates breaks at a small proportion of one or two of the four nucleotide bases in each of four reactions (G, A+G, C, C+T). The concentration of the modifying chemicals is controlled to introduce on average one modification per DNA molecule. Thus a series of labeled fragments is generated, from the radiolabeled end to the first "cut" site in each molecule. The fragments in the four reactions are electrophoresed side by side in denaturing acrylamide gels for size separation. To visualize the fragments, the gel is exposed to X-ray film for autoradiography, yielding a series of dark bands each corresponding to a radiolabeled DNA fragment, from which the sequence may be inferred.

Chain-termination Methods

The chain-termination method developed by Frederick Sanger and coworkers in 1977 soon became the method of choice, owing to its relative ease and reliability. When invented, the chain-terminator method used fewer toxic chemicals and lower amounts of radioactivity than the Maxam and Gilbert method. Because of its comparative ease, the Sanger method was soon automated and was the method used in the first generation of DNA sequencers.

Sanger sequencing is the method which prevailed from the 1980s until the mid-2000s. Over that period, great advances were made in the technique, such as fluorescent labelling, capillary electrophoresis, and general automation. These developments allowed much more efficient sequencing, leading to lower costs. The Sanger method, in mass production form, is the technology which produced the first human genome in 2001, ushering in the age of genomics. However, later in the decade, radically different approaches reached the market, bringing the cost per genome down from $100 million in 2001 to $10,000 in 2011.

Advanced Methods and de Novo Sequencing

Large-scale sequencing often aims at sequencing very long DNA pieces, such as whole chromosomes, although large-scale sequencing can also be used to generate very large numbers of short sequences, such as found in phage display. For longer targets such as chromosomes, common approaches consist of cutting (with restriction enzymes) or shearing (with mechanical forces) large DNA fragments into shorter DNA fragments. The fragmented DNA may then be cloned into a DNA vector and amplified in a bacterial host such as *Escherichia coli*. Short DNA fragments purified from individual bacterial colonies are individually sequenced and assembled electronically into one long, contiguous sequence. Studies have shown that adding a size selection step to collect DNA fragments of uniform size can improve sequencing efficiency and accuracy of the genome assembly. In these studies, automated sizing has proven to be more reproducible and precise than manual gel sizing.

The term "*de novo* sequencing" specifically refers to methods used to determine the sequence of DNA with no previously known sequence. *De novo* translates from Latin as "from the beginning".

Gaps in the assembled sequence may be filled by primer walking. The different strategies have different tradeoffs in speed and accuracy; shotgun methods are often used for sequencing large genomes, but its assembly is complex and difficult, particularly with sequence repeats often causing gaps in genome assembly.

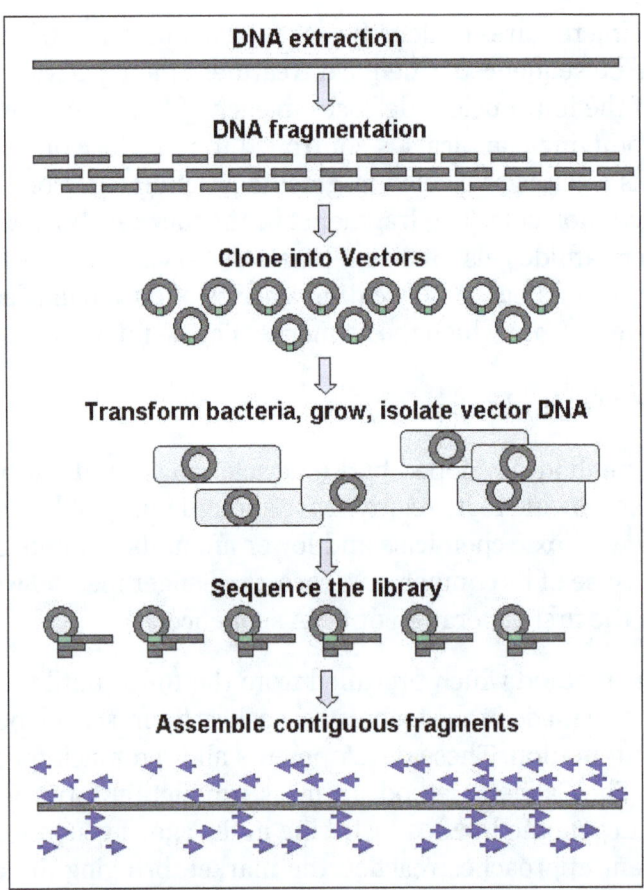

Genomic DNA is fragmented into random pieces and cloned as a bacterial library. DNA from individual bacterial clones is sequenced and the sequence is assembled by using overlapping DNA regions.

Most sequencing approaches use an *in vitro* cloning step to amplify individual DNA molecules, because their molecular detection methods are not sensitive enough for single molecule sequencing. Emulsion PCR isolates individual DNA molecules along with primer-coated beads in aqueous droplets within an oil phase. A polymerase chain reaction (PCR) then coats each bead with clonal copies of the DNA molecule followed by immobilization for later sequencing. Emulsion PCR is used in the methods developed by Marguilis et al., Shendure and Porreca et al. (also known as "Polony sequencing") and SOLiD sequencing. Emulsion PCR is also used in the GemCode and Chromium platforms developed by 10× Genomics.

Shotgun Sequencing

Shotgun sequencing is a sequencing method designed for analysis of DNA sequences longer than 1000 base pairs, up to and including entire chromosomes. This method requires the target DNA to be broken into random fragments. After sequencing individual fragments, the sequences can be reassembled on the basis of their overlapping regions.

Bridge PCR

Another method for *in vitro* clonal amplification is bridge PCR, in which fragments are amplified upon primers attached to a solid surface and form "DNA colonies" or "DNA clusters". This method is used in the Illumina Genome Analyzer sequencers. Single-molecule methods, such as that developed by Stephen Quake's laboratory are an exception: they use bright fluorophores and laser excitation to detect base addition events from individual DNA molecules fixed to a surface, eliminating the need for molecular amplification.

High-throughput Methods

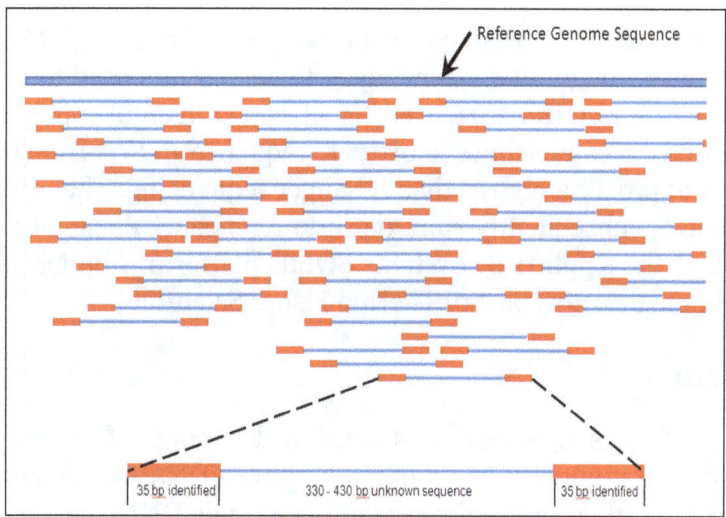

Multiple, fragmented sequence reads must be assembled together on the basis of their overlapping areas.

High-throughput, or next-generation, sequencing applies to genome sequencing, genome resequencing, transcriptome profiling (RNA-Seq), DNA-protein interactions (ChIP-sequencing), and epigenome characterization. Resequencing is necessary, because the genome of a single individual of a species will not indicate all of the genome variations among other individuals of the same species.

The high demand for low-cost sequencing has driven the development of high-throughput sequencing technologies that parallelize the sequencing process, producing thousands or millions of sequences concurrently. High-throughput sequencing technologies are intended to lower the cost of DNA sequencing beyond what is possible with standard dye-terminator methods. In ultra-high-throughput sequencing as many as 500,000 sequencing-by-synthesis operations may be run in parallel. Such technologies led to the ability to sequence an entire human genome in as little as one day.

Massively Parallel Signature Sequencing

The first of the high-throughput sequencing technologies, massively parallel signature sequencing (or MPSS), was developed in the 1990s at Lynx Therapeutics, a company founded in 1992 by Sydney Brenner and Sam Eletr. MPSS was a bead-based method that used a complex approach of adapter ligation followed by adapter decoding, reading the sequence in increments of four nucleotides.

This method made it susceptible to sequence-specific bias or loss of specific sequences. Because the technology was so complex, MPSS was only performed 'in-house' by Lynx Therapeutics and no DNA sequencing machines were sold to independent laboratories. Lynx Therapeutics merged with Solexa (later acquired by Illumina) in 2004, leading to the development of sequencing-by-synthesis, a simpler approach acquired from Manteia Predictive Medicine, which rendered MPSS obsolete. However, the essential properties of the MPSS output were typical of later high-throughput data types, including hundreds of thousands of short DNA sequences. In the case of MPSS, these were typically used for sequencing cDNA for measurements of gene expression levels.

Polony Sequencing

The Polony sequencing method, developed in the laboratory of George M. Church at Harvard, was among the first high-throughput sequencing systems and was used to sequence a full *E. coli* genome in 2005. It combined an in vitro paired-tag library with emulsion PCR, an automated microscope, and ligation-based sequencing chemistry to sequence an *E. coli* genome at an accuracy of >99.9999% and a cost approximately 1/9 that of Sanger sequencing. The technology was licensed to Agencourt Biosciences, subsequently spun out into Agencourt Personal Genomics, and eventually incorporated into the Applied Biosystems SOLiD platform. Applied Biosystems was later acquired by Life Technologies, now part of Thermo Fisher Scientific.

454 Pyrosequencing

A parallelized version of pyrosequencing was developed by 454 Life Sciences, which has since been acquired by Roche Diagnostics. The method amplifies DNA inside water droplets in an oil solution (emulsion PCR), with each droplet containing a single DNA template attached to a single primer-coated bead that then forms a clonal colony. The sequencing machine contains many picoliter-volume wells each containing a single bead and sequencing enzymes. Pyrosequencing uses luciferase to generate light for detection of the individual nucleotides added to the nascent DNA, and the combined data are used to generate sequence reads. This technology provides intermediate read length and price per base compared to Sanger sequencing on one end and Solexa and SOLiD on the other.

Illumina Sequencing

An Illumina HiSeq 2500 sequencer.

Solexa, now part of Illumina, was founded by Shankar Balasubramanian and David Klenerman in 1998, and developed a sequencing method based on reversible dye-terminators technology, and engineered polymerases. The reversible terminated chemistry concept was invented by Bruno Canard and Simon Sarfati at the Pasteur Institute in Paris. It was developed internally at Solexa by those named on the relevant patents. In 2004, Solexa acquired the company Manteia Predictive Medicine in order to gain a massively parallel sequencing technology invented in 1997 by Pascal Mayer and Laurent Farinelli. It is based on "DNA Clusters" or "DNA colonies", which involves the clonal amplification of DNA on a surface.

In this method, DNA molecules and primers are first attached on a slide or flow cell and amplified with polymerase so that local clonal DNA colonies, later coined "DNA clusters", are formed. To determine the sequence, four types of reversible terminator bases (RT-bases) are added and non-incorporated nucleotides are washed away. A camera takes images of the fluorescently labeled nucleotides. Then the dye, along with the terminal 3' blocker, is chemically removed from the DNA, allowing for the next cycle to begin. Unlike pyrosequencing, the DNA chains are extended one nucleotide at a time and image acquisition can be performed at a delayed moment, allowing for very large arrays of DNA colonies to be captured by sequential images taken from a single camera.

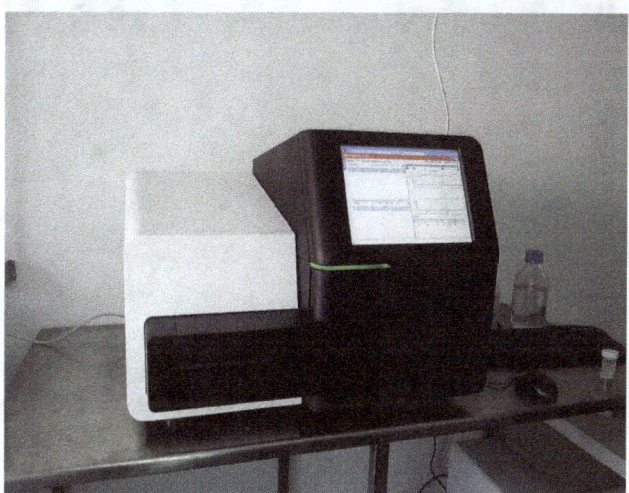
An Illumina MiSeq sequencer.

Decoupling the enzymatic reaction and the image capture allows for optimal throughput and theoretically unlimited sequencing capacity. With an optimal configuration, the ultimately reachable instrument throughput is thus dictated solely by the analog-to-digital conversion rate of the camera, multiplied by the number of cameras and divided by the number of pixels per DNA colony required for visualizing them optimally (approximately 10 pixels/colony). In 2012, with cameras operating at more than 10 MHz A/D conversion rates and available optics, fluidics and enzymatics, throughput can be multiples of 1 million nucleotides/second, corresponding roughly to 1 human genome equivalent at 1× coverage per hour per instrument, and 1 human genome re-sequenced (at approx. 30×) per day per instrument (equipped with a single camera).

Combinatorial Probe Anchor Synthesis

This method is an upgraded modification to combinatorial probe anchor ligation technology (cPAL) described by Complete Genomics which has since become part of Chinese genomics company BGI in 2013. The two companies have refined the technology to allow for longer read lengths, reaction time reductions and faster time to results. In addition, data are now generated as contiguous full-length reads in the standard FASTQ file format and can be used as-is in most short-read-based bioinformatics analysis pipelines.

The two technologies that form the basis for this high-throughput sequencing technology are DNA nanoballs (DNB) and patterned arrays for nanoball attachment to a solid surface. DNA nanoballs are simply formed by denaturing double stranded, adapter ligated libraries and ligating the forward strand only to a splint oligonucleotide to form a ssDNA circle. Faithful copies of the circles containing the DNA insert are produced utilizing Rolling Circle Amplification that generates approximately 300–500 copies. The long strand of ssDNA folds upon itself to produce a three-dimensional nanoball structure that is approximately 220 nm in diameter. Making DNBs replaces the need to generate PCR copies of the library on the flow cell and as such can remove large proportions of duplicate reads, adapter-adapter ligations and PCR induced errors.

A BGI MGISEQ-2000RS sequencer.

The patterned array of positively charged spots is fabricated through photolithography and etching techniques followed by chemical modification to generate a sequencing flow cell. Each spot on the flow cell is approximately 250 nm in diameter, are separated by 700 nm (centre to centre) and allows easy attachment of a single negatively charged DNB to the flow cell and thus reducing under or over-clustering on the flow cell.

Sequencing is then performed by addition of an oligonucleotide probe that attaches in combination to specific sites within the DNB. The probe acts as an anchor that then allows one of four single reversibly inactivated, labelled nucleotides to bind after flowing across the flow cell. Unbound nucleotides are washed away before laser excitation of the attached labels then emit fluorescence and signal is captured by cameras that is converted to a digital output for base calling. The attached base has its terminator and label chemically cleaved at completion of the cycle. The cycle is repeated with another flow of free, labelled nucleotides across the flow cell to allow the next nucleotide to bind and have its signal captured. This process is completed a number of times (usually 50 to 300 times) to determine the sequence of the inserted piece of DNA at a rate of approximately 40 million nucleotides per second as of 2018.

SOLiD Sequencing

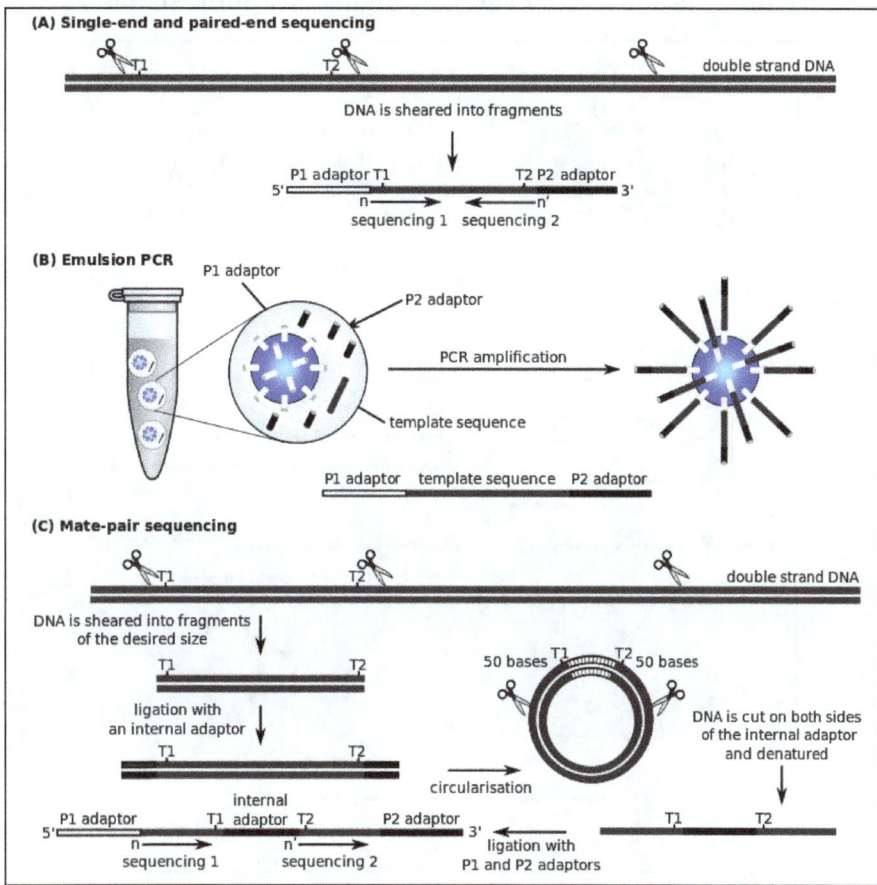

Library preparation for the SOLiD platform.

Applied Biosystems' SOLiD technology employs sequencing by ligation. Here, a pool of all possible oligonucleotides of a fixed length are labeled according to the sequenced position. Oligonucleotides are annealed and ligated; the preferential ligation by DNA ligase for matching sequences results in a signal informative of the nucleotide at that position. Before sequencing, the DNA is amplified by emulsion PCR. The resulting beads, each containing single copies of the same DNA molecule, are deposited on a glass slide. The result is sequences of quantities and lengths comparable to Illumina sequencing. This sequencing by ligation method has been reported to have some issue sequencing palindromic sequences.

Ion Torrent Semiconductor Sequencing

Ion Torrent Systems developed a system based on using standard sequencing chemistry, but with a novel, semiconductor-based detection system. This method of sequencing is based on the detection of hydrogen ions that are released during the polymerisation of DNA, as opposed to the optical methods used in other sequencing systems. A microwell containing a template DNA strand to be sequenced is flooded with a single type of nucleotide. If the introduced nucleotide is complementary to the leading template nucleotide it is incorporated into the growing complementary strand. This causes the release of a hydrogen ion that triggers a hypersensitive ion sensor, which indicates that a reaction has occurred. If homopolymer repeats are present in the template sequence, multiple nucleotides will be incorporated in a single cycle. This leads to a corresponding number of released hydrogens and a proportionally higher electronic signal.

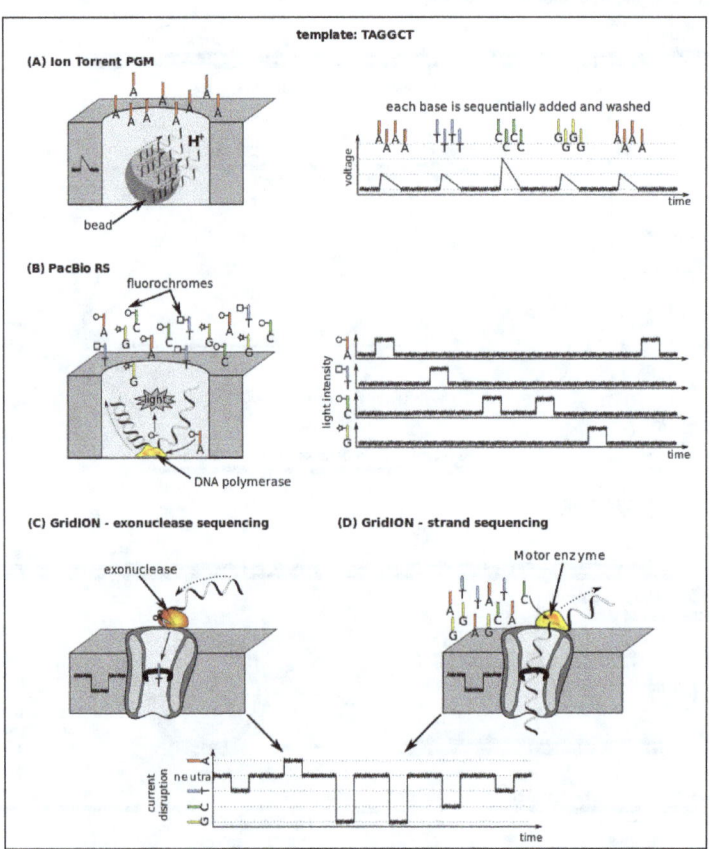

Sequencing of the TAGGCT template with IonTorrent, PacBioRS and Grid ION.

DNA Nanoball Sequencing

DNA nanoball sequencing is a type of high throughput sequencing technology used to determine the entire genomic sequence of an organism. The company Complete Genomics uses this technology to sequence samples submitted by independent researchers. The method uses rolling circle replication to amplify small fragments of genomic DNA into DNA nanoballs. Unchained sequencing by ligation is then used to determine the nucleotide sequence. This method of DNA sequencing allows large numbers of DNA nanoballs to be sequenced per run and at low reagent costs compared to other high-throughput sequencing platforms. However, only short sequences of DNA are determined from each DNA nanoball which makes mapping the short reads to a reference genome difficult.

Heliscope Single Molecule Sequencing

Heliscope sequencing is a method of single-molecule sequencing developed by Helicos Biosciences. It uses DNA fragments with added poly-A tail adapters which are attached to the flow cell surface. The next steps involve extension-based sequencing with cyclic washes of the flow cell with fluorescently labeled nucleotides (one nucleotide type at a time, as with the Sanger method). The reads are performed by the Heliscope sequencer. The reads are short, averaging 35 bp. In 2009 a human genome was sequenced using the Heliscope.

Single Molecule Real Time Sequencing

Single Molecule Real Time (SMRT) sequencing is based on the sequencing by synthesis approach. The DNA is synthesized in zero-mode wave-guides (ZMWs) – small well-like containers with the capturing tools located at the bottom of the well. The sequencing is performed with use of unmodified polymerase (attached to the ZMW bottom) and fluorescently labelled nucleotides flowing freely in the solution. The wells are constructed in a way that only the fluorescence occurring by the bottom of the well is detected. The fluorescent label is detached from the nucleotide upon its incorporation into the DNA strand, leaving an unmodified DNA strand. According to Pacific Biosciences (PacBio), the SMRT technology developer, this methodology allows detection of nucleotide modifications (such as cytosine methylation). This happens through the observation of polymerase kinetics. This approach allows reads of 20,000 nucleotides or more, with average read lengths of 5 kilobases. In 2015, Pacific Biosciences announced the launch of a new sequencing instrument called the Sequel System, with 1 million ZMWs compared to 150,000 ZMWs in the PacBio RS II instrument. SMRT sequencing is referred to as "third-generation" or "long-read" sequencing.

Nanopore DNA Sequencing

The DNA passing through the nanopore changes its ion current. This change is dependent on the shape, size and length of the DNA sequence. Each type of the nucleotide blocks the ion flow through the pore for a different period of time. The method does not require modified nucleotides and is performed in real time. Nanopore sequencing is referred to as "third-generation" or "long-read" sequencing, along with SMRT sequencing.

Early industrial research into this method was based on a technique called 'Exonuclease sequencing', where the readout of electrical signals occurring at nucleotides passing by alpha(α)-hemolysin pores covalently bound with cyclodextrin. However the subsequently commercial method, 'strand sequencing' sequencing DNA bases in an intact strand.

Two main areas of nanopore sequencing in development are solid state nanopore sequencing, and protein based nanopore sequencing. Protein nanopore sequencing utilizes membrane protein complexes such as α-hemolysin, MspA (*Mycobacterium smegmatis* Porin A) or CssG, which show great promise given their ability to distinguish between individual and groups of nucleotides. In contrast, solid-state nanopore sequencing utilizes synthetic materials such as silicon nitride and aluminum oxide and it is preferred for its superior mechanical ability and thermal and chemical stability. The fabrication method is essential for this type of sequencing given that the nanopore array can contain hundreds of pores with diameters smaller than eight nanometers.

The concept originated from the idea that single stranded DNA or RNA molecules can be electrophoretically driven in a strict linear sequence through a biological pore that can be less than eight nanometers, and can be detected given that the molecules release an ionic current while moving through the pore. The pore contains a detection region capable of recognizing different bases, with each base generating various time specific signals corresponding to the sequence of bases as they cross the pore which are then evaluated. Precise control over the DNA transport through the pore is crucial for success. Various enzymes such as exonucleases and polymerases have been used to moderate this process by positioning them near the pore's entrance.

Microfluidic Systems

There are two main microfluidic systems that are used to sequence DNA; droplet based microfluidics and digital microfluidics. Microfluidic devices solve many of the current limitations of current sequencing arrays.

Abate et al. studied the use of droplet-based microfluidic devices for DNA sequencing. These devices have the ability to form and process picoliter sized droplets at the rate of thousands per second. The devices were created from polydimethylsiloxane (PDMS) and used Forster resonance energy transfer, FRET assays to read the sequences of DNA encompassed in the droplets. Each position on the array tested for a specific 15 base sequence.

Fair et al. used digital microfluidic devices to study DNA pyrosequencing. Significant advantages include the portability of the device, reagent volume, speed of analysis, mass manufacturing abilities, and high throughput. This study provided a proof of concept showing that digital devices can be used for pyrosequencing; the study included using synthesis, which involves the extension of the enzymes and addition of labeled nucleotides.

Boles et al. also studied pyrosequencing on digital microfluidic devices. They used an electro-wetting device to create, mix, and split droplets. The sequencing uses a three-enzyme protocol and DNA templates anchored with magnetic beads. The device was tested using two protocols and resulted in 100% accuracy based on raw pyrogram levels. The advantages of these digital microfluidic devices include size, cost, and achievable levels of functional integration.

DNA sequencing research, using microfluidics, also has the ability to be applied to the sequencing of RNA, using similar droplet microfluidic techniques, such as the method, inDrops. This shows that many of these DNA sequencing techniques will be able to be applied further and be used to understand more about genomes and transcriptomes.

Methods in Development

DNA sequencing methods currently under development include reading the sequence as a DNA strand transits through nanopores (a method that is now commercial but subsequent generations such as solid-state nanopores are still in development), and microscopy-based techniques, such as atomic force microscopy or transmission electron microscopy that are used to identify the positions of individual nucleotides within long DNA fragments (>5,000 bp) by nucleotide labeling with heavier elements (e.g., halogens) for visual detection and recording. Third generation technologies aim to increase throughput and decrease the time to result and cost by eliminating the need for excessive reagents and harnessing the processivity of DNA polymerase.

Tunnelling Currents DNA Sequencing

Another approach uses measurements of the electrical tunnelling currents across single-strand DNA as it moves through a channel. Depending on its electronic structure, each base affects the tunnelling current differently, allowing differentiation between different bases.

The use of tunnelling currents has the potential to sequence orders of magnitude faster than ionic current methods and the sequencing of several DNA oligomers and micro-RNA has already been achieved.

Sequencing by Hybridization

Sequencing by hybridization is a non-enzymatic method that uses a DNA microarray. A single pool of DNA whose sequence is to be determined is fluorescently labeled and hybridized to an array containing known sequences. Strong hybridization signals from a given spot on the array identifies its sequence in the DNA being sequenced.

This method of sequencing utilizes binding characteristics of a library of short single stranded DNA molecules (oligonucleotides), also called DNA probes, to reconstruct a target DNA sequence. Non-specific hybrids are removed by washing and the target DNA is eluted. Hybrids are re-arranged such that the DNA sequence can be reconstructed. The benefit of this sequencing type is its ability to capture a large number of targets with a homogenous coverage. A large number of chemicals and starting DNA is usually required. However, with the advent of solution-based hybridization, much less equipment and chemicals are necessary.

Sequencing with Mass Spectrometry

Mass spectrometry may be used to determine DNA sequences. Matrix-assisted laser desorption ionization time-of-flight mass spectrometry, or MALDI-TOF MS, has specifically been investigated as an alternative method to gel electrophoresis for visualizing DNA fragments. With this method, DNA fragments generated by chain-termination sequencing reactions are compared by mass rather than by size. The mass of each nucleotide is different from the others and this difference is detectable by mass spectrometry. Single-nucleotide mutations in a fragment can be more easily detected with MS than by gel electrophoresis alone. MALDI-TOF MS can more easily detect differences between RNA fragments, so researchers may indirectly sequence DNA with MS-based methods by converting it to RNA first.

The higher resolution of DNA fragments permitted by MS-based methods is of special interest to researchers in forensic science, as they may wish to find single-nucleotide polymorphisms in human DNA samples to identify individuals. These samples may be highly degraded so forensic researchers often prefer mitochondrial DNA for its higher stability and applications for lineage studies. MS-based sequencing methods have been used to compare the sequences of human mitochondrial DNA from samples in a Federal Bureau of Investigation database and from bones found in mass graves of World War I soldiers.

Early chain-termination and TOF MS methods demonstrated read lengths of up to 100 base pairs. Researchers have been unable to exceed this average read size; like chain-termination sequencing alone, MS-based DNA sequencing may not be suitable for large *de novo* sequencing projects. Even so, a recent study did use the short sequence reads and mass spectroscopy to compare single-nucleotide polymorphisms in pathogenic *Streptococcus* strains.

Microfluidic Sanger Sequencing

In microfluidic Sanger sequencing the entire thermocycling amplification of DNA fragments as well as their separation by electrophoresis is done on a single glass wafer (approximately 10 cm in diameter) thus reducing the reagent usage as well as cost. In some instances researchers have shown that they can increase the throughput of conventional sequencing through the use of microchips.

Microscopy-based Techniques

This approach directly visualizes the sequence of DNA molecules using electron microscopy. The first identification of DNA base pairs within intact DNA molecules by enzymatically incorporating modified bases, which contain atoms of increased atomic number, direct visualization and identification of individually labeled bases within a synthetic 3,272 base-pair DNA molecule and a 7,249 base-pair viral genome has been demonstrated.

RNAP Sequencing

This method is based on use of RNA polymerase (RNAP), which is attached to a polystyrene bead. One end of DNA to be sequenced is attached to another bead, with both beads being placed in optical traps. RNAP motion during transcription brings the beads in closer and their relative distance changes, which can then be recorded at a single nucleotide resolution. The sequence is deduced based on the four readouts with lowered concentrations of each of the four nucleotide types, similarly to the Sanger method. A comparison is made between regions and sequence information is deduced by comparing the known sequence regions to the unknown sequence regions.

In Vitro Virus High-throughput Sequencing

A method has been developed to analyze full sets of protein interactions using a combination of 454 pyrosequencing and an *in vitro* virus mRNA display method. Specifically, this method covalently links proteins of interest to the mRNAs encoding them, then detects the mRNA pieces using reverse transcription PCRs. The mRNA may then be amplified and sequenced. The combined method was titled IVV-HiTSeq and can be performed under cell-free conditions, though its results may not be representative of *in vivo* conditions.

Sample Preparation

The success of any DNA sequencing protocol relies upon the DNA or RNA sample extraction and preparation from the biological material of interest.

- A successful DNA extraction will yield a DNA sample with long, non-degraded strands.
- A successful RNA extraction will yield a RNA sample that should be converted to complementary DNA (cDNA) using reverse transcriptase—a DNA polymerase that synthesizes a complementary DNA based on existing strands of RNA in a PCR-like manner. Complementary DNA can then be processed the same way as genomic DNA.

According to the sequencing technology to be used, the samples resulting from either the DNA or the RNA extraction require further preparation. For Sanger sequencing, either cloning procedures or PCR are required prior to sequencing. In the case of next-generation sequencing methods, library preparation is required before processing. Assessing the quality and quantity of nucleic acids both after extraction and after library preparation identifies degraded, fragmented, and low-purity samples and yields high-quality sequencing data.

Computational Challenges

The sequencing technologies described here produce raw data that needs to be assembled into longer sequences such as complete genomes (sequence assembly). There are many computational challenges to achieve this, such as the evaluation of the raw sequence data which is done by programs and algorithms such as Phred and Phrap. Other challenges have to deal with repetitive sequences that often prevent complete genome assemblies because they occur in many places of the genome. As a consequence, many sequences may not be assigned to particular chromosomes. The production of raw sequence data is only the beginning of its detailed bioinformatical analysis. Yet new methods for sequencing and correcting sequencing errors were developed.

Read Trimming

Sometimes, the raw reads produced by the sequencer are correct and precise only in a fraction of their length. Using the entire read may introduce artifacts in the downstream analyses like genome assembly, snp calling, or gene expression estimation. Two classes of trimming programs have been introduced, based on the window-based or the running-sum classes of algorithms. This is a partial list of the trimming algorithms currently available, specifying the algorithm class they belong to:

Read Trimming Algorithms	
Name of algorithm	Type of algorithm
Cutadapt	Running sum
ConDeTri	Window based
ERNE-FILTER	Running sum
FASTX quality trimmer	Window based
PRINSEQ	Window based
Trimmomatic	Window based
SolexaQA	Window based
SolexaQA-BWA	Running sum
Sickle	Window based

Sanger Sequencing

Sanger sequencing is a method of DNA sequencing first commercialized by Applied Biosystems, based on the selective incorporation of chain-terminating dideoxynucleotides by DNA polymerase during in vitro DNA replication. Developed by Frederick Sanger and colleagues in 1977, it was the most widely used sequencing method for approximately 40 years. More recently, higher volume Sanger sequencing has been replaced by "Next-Gen" sequencing methods, especially for large-scale, automated genome analyses. However, the Sanger method remains in wide use, for smaller-scale projects, and for validation of Next-Gen results. It still has the advantage over short-read sequencing technologies (like Illumina) that it can produce DNA sequence reads of > 500 nucleotides.

Method

The classical chain-termination method requires a single-stranded DNA template, a DNA primer, a DNA polymerase, normal deoxynucleotidetriphosphates (dNTPs), and modified di-deoxynucleotidetriphosphates (ddNTPs), the latter of which terminate DNA strand elongation. These chain-terminating nucleotides lack a 3'-OH group required for the formation of a phosphodiester bond between two nucleotides, causing DNA polymerase to cease extension of DNA when a modified ddNTP is incorporated. The ddNTPs may be radioactively or fluorescently labelled for detection in automated sequencing machines.

The DNA sample is divided into four separate sequencing reactions, containing all four of the standard deoxynucleotides (dATP, dGTP, dCTP and dTTP) and the DNA polymerase. To each reaction is added only one of the four dideoxynucleotides (ddATP, ddGTP, ddCTP, or ddTTP), while the other added nucleotides are ordinary ones. The dideoxynucleotide concentration should be approximately 100-fold lower than that of the corresponding deoxynucleotide (e.g. 0.005 mM ddTTP: 0.5 mM dTTP) to allow enough fragments to be produced while still transcribing the complete sequence. Putting it in a more sensible order, four separate reactions are needed in this process to test all four ddNTPs. Following rounds of template DNA extension from the bound primer, the resulting DNA fragments are heat denatured and separated by size using gel electrophoresis. In the original publication of 1977, the formation of base-paired loops of ssDNA was a cause of serious difficulty in resolving bands at some locations. This is frequently performed using a denaturing polyacrylamide-urea gel with each of the four reactions run in one of four individual lanes (lanes A, T, G, C). The DNA bands may then be visualized by autoradiography or UV light and the DNA sequence can be directly read off the X-ray film or gel image.

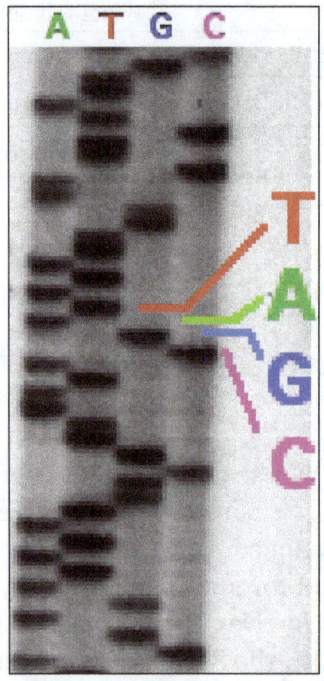

Part of a radioactively labelled sequencing gel.

In the image, X-ray film was exposed to the gel, and the dark bands correspond to DNA fragments of different lengths. A dark band in a lane indicates a DNA fragment that is the result of chain termination after incorporation of a dideoxynucleotide (ddATP, ddGTP, ddCTP, or ddTTP). The relative positions of the different bands among the four lanes, from bottom to top, are then used to read the DNA sequence.

Technical variations of chain-termination sequencing include tagging with nucleotides containing radioactive phosphorus for radiolabelling, or using a primer labeled at the 5' end with a fluorescent dye. Dye-primer sequencing facilitates reading in an optical system for faster and more economical analysis and automation. The later development by Leroy Hood and coworkers of fluorescently labeled ddNTPs and primers set the stage for automated, high-throughput DNA sequencing.

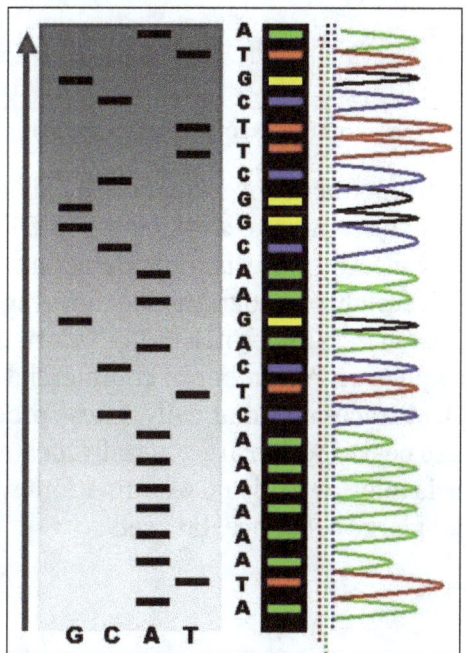

Sequence ladder by radioactive sequencing compared to fluorescent peaks.

Chain-termination methods have greatly simplified DNA sequencing. For example, chain-termination-based kits are commercially available that contain the reagents needed for sequencing, pre-aliquoted and ready to use. Limitations include non-specific binding of the primer to the DNA, affecting accurate read-out of the DNA sequence, and DNA secondary structures affecting the fidelity of the sequence.

Dye-terminator Sequencing

Dye-terminator sequencing utilizes labelling of the chain terminator ddNTPs, which permits sequencing in a single reaction, rather than four reactions as in the labelled-primer method. In dye-terminator sequencing, each of the four dideoxynucleotide chain terminators is labelled with fluorescent dyes, each of which emit light at different wavelengths.

Owing to its greater expediency and speed, dye-terminator sequencing is now the mainstay in automated sequencing. Its limitations include dye effects due to differences in the incorporation of the dye-labelled chain terminators into the DNA fragment, resulting in unequal peak heights and shapes in the electronic DNA sequence trace chromatogram after capillary electrophoresis.

This problem has been addressed with the use of modified DNA polymerase enzyme systems and dyes that minimize incorporation variability, as well as methods for eliminating "dye blobs". The dye-terminator sequencing method, along with automated high-throughput DNA sequence analyzers, was used for the vast majority of sequencing projects until the introduction of Next Generation Sequencing.

Automation and Sample Preparation

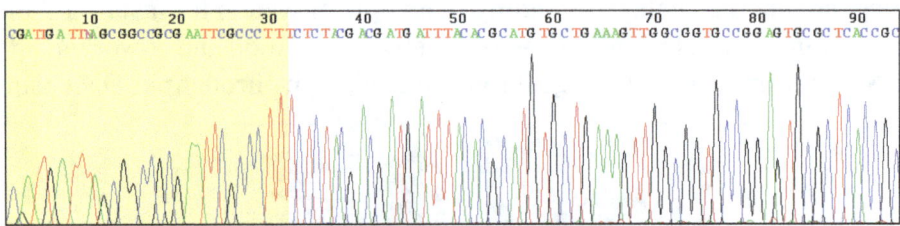

View of the start of an example dye-terminator read.

Automated DNA-sequencing instruments (DNA sequencers) can sequence up to 384 DNA samples in a single batch. Batch runs may occur up to 24 times a day. DNA sequencers separate strands by size (or length) using capillary electrophoresis, they detect and record dye fluorescence, and output data as fluorescent peak trace chromatograms. Sequencing reactions (thermocycling and labelling), cleanup and re-suspension of samples in a buffer solution are performed separately, before loading samples onto the sequencer. A number of commercial and non-commercial software packages can trim low-quality DNA traces automatically. These programs score the quality of each peak and remove low-quality base peaks (which are generally located at the ends of the sequence). The accuracy of such algorithms is inferior to visual examination by a human operator, but is adequate for automated processing of large sequence data sets.

Challenges

Common challenges of DNA sequencing with the Sanger method include poor quality in the first 15-40 bases of the sequence due to primer binding and deteriorating quality of sequencing traces after 700-900 bases. Base calling software such as Phred typically provides an estimate of quality to aid in trimming of low-quality regions of sequences.

In cases where DNA fragments are cloned before sequencing, the resulting sequence may contain parts of the cloning vector. In contrast, PCR-based cloning and next-generation sequencing technologies based on pyrosequencing often avoid using cloning vectors. Recently, one-step Sanger sequencing (combined amplification and sequencing) methods such as Ampliseq and SeqSharp have been developed that allow rapid sequencing of target genes without cloning or prior amplification.

Current methods can directly sequence only relatively short (300-1000 nucleotides long) DNA fragments in a single reaction. The main obstacle to sequencing DNA fragments above this size limit is insufficient power of separation for resolving large DNA fragments that differ in length by only one nucleotide.

Microfluidic Sanger Sequencing

Microfluidic Sanger sequencing is a lab-on-a-chip application for DNA sequencing, in which the Sanger sequencing steps (thermal cycling, sample purification, and capillary electrophoresis) are integrated on a wafer-scale chip using nanoliter-scale sample volumes. This technology generates long and accurate sequence reads, while obviating many of the significant shortcomings of the conventional Sanger method (e.g. high consumption of expensive reagents, reliance on expensive equipment, personnel-intensive manipulations, etc.) by integrating and automating the Sanger sequencing steps.

In its modern inception, high-throughput genome sequencing involves fragmenting the genome into small single-stranded pieces, followed by amplification of the fragments by Polymerase Chain Reaction (PCR). Adopting the Sanger method, each DNA fragment is irreversibly terminated with the incorporation of a fluorescently labeled dideoxy chain-terminating nucleotide, thereby producing a DNA "ladder" of fragments that each differ in length by one base and bear a base-specific fluorescent label at the terminal base. Amplified base ladders are then separated by Capillary Array Electrophoresis (CAE) with automated, *in situ* "finish-line" detection of the fluorescently labeled ssDNA fragments, which provides an ordered sequence of the fragments. These sequence reads are then computer assembled into overlapping or contiguous sequences (termed "contigs") which resemble the full genomic sequence once fully assembled.

Sanger methods achieve read lengths of approximately 800bp (typically 500-600 bp with non-enriched DNA). The longer read lengths in Sanger methods display significant advantages over other sequencing methods especially in terms of sequencing repetitive regions of the genome. A challenge of short-read sequence data is particularly an issue in sequencing new genomes *(de novo)* and in sequencing highly rearranged genome segments, typically those seen of cancer genomes or in regions of chromosomes that exhibit structural variation.

Applications of Microfluidic Sequencing Technologies

Other useful applications of DNA sequencing include single nucleotide polymorphism (SNP) detection, single-strand conformation polymorphism (SSCP) heteroduplex analysis, and short tandem repeat (STR) analysis. Resolving DNA fragments according to differences in size and conformation is the most critical step in studying these features of the genome.

Device Design

The sequencing chip has a four-layer construction, consisting of three 100 mm-diameter glass wafers (on which device elements are microfabricated) and a polydimethylsiloxane (PDMS) membrane. Reaction chambers and capillary electrophoresis channels are etched between the top two glass wafers, which are thermally bonded. Three-dimensional channel interconnections and microvalves are formed by the PDMS and bottom manifold glass wafer.

The device consists of three functional units, each corresponding to the Sanger sequencing steps. The Thermal Cycling (TC) unit is a 250-nanoliter reaction chamber with integrated resistive temperature detector, microvalves, and a surface heater. Movement of reagent between the top all-glass layer and the lower glass-PDMS layer occurs through 500 μm-diameter via-holes. After thermal-cycling, the reaction mixture undergoes purification in the capture/purification chamber, and then is injected into the capillary electrophoresis (CE) chamber. The CE unit consists of a 30 cm capillary which is folded into a compact switchback pattern via 65 μm-wide turns.

Sequencing Chemistry

Thermal Cycling

In the TC reaction chamber, dye-terminator sequencing reagent, template DNA, and primers are loaded into the TC chamber and thermal-cycled for 35 cycles (at 95 °C for 12 seconds and at 60 °C for 55 seconds).

Purification

The charged reaction mixture (containing extension fragments, template DNA, and excess sequencing reagent) is conducted through a capture/purification chamber at 30 °C via a 33-Volts/cm electric field applied between capture outlet and inlet ports. The capture gel through which the sample is driven, consists of 40 µM of oligonucleotide (complementary to the primers) covalently bound to a polyacrylamide matrix. Extension fragments are immobilized by the gel matrix, and excess primer, template, free nucleotides, and salts are eluted through the capture waste port. The capture gel is heated to 67-75 °C to release extension fragments.

Capillary Electrophoresis

Extension fragments are injected into the CE chamber where they are electrophoresed through a 125-167-V/cm field.

Platforms

The Apollo 100 platform integrates the first two Sanger sequencing steps (thermal cycling and purification) in a fully automated system. The manufacturer claims that samples are ready for capillary electrophoresis within three hours of the sample and reagents being loaded into the system. The Apollo 100 platform requires sub-microliter volumes of reagents.

Comparisons to Other Sequencing Techniques

Table: Performance values for genome sequencing technologies including Sanger methods and next-generation methods.

Technology	Number of lanes	Injection volume (nL)	Analysis time	Average read length	Throughput (including analysis; Mb/h)	Gel pouring	Lane tracking
Slab gel	96	500–1000	6–8 hours	700 bp	0.0672	Yes	Yes
Capillary array electrophoresis	96	1–5	1–3 hours	700 bp	0.166	No	No
Microchip	96	0.1–0.5	6–30 minutes	430 bp	0.660	No	No
454/Roche FLX		< 0.001	4 hours	200–300 bp	20–30	No	
Illumina/Solexa			2–3 days	30–100 bp	20	No	
ABI/SOLiD			8 days	35 bp	5–15	No	
Illumina MiSeq			1–3 days	2×75–2×300 bp	170–250	No	
Illumina Nova-Seq			1–2 days	2×50–2×150 bp	22,000–67,000	No	
Ion Torrent Ion 530			2.5–4 hours	200–600 bp	110–920	No	
BGI MGISEQ-T7			1 day	2×150 bp	250,000	No	
Pacific Biosciences			10–20 hours	10–30 kb	1,300	No	
Oxford Nanopore MinIon			3 days	13–20 kb	700	No	

The ultimate goal of high-throughput sequencing is to develop systems that are low-cost, and extremely efficient at obtaining extended (longer) read lengths. Longer read lengths of each single electrophoretic separation, substantially reduces the cost associated with de novo DNA sequencing and the number of templates needed to sequence DNA contigs at a given redundancy. Microfluidics may allow for faster, cheaper and easier sequence assembly.

Ion Semiconductor Sequencing

Ion semiconductor sequencing is a method of DNA sequencing based on the detection of hydrogen ions that are released during the polymerization of DNA. This is a method of "sequencing by synthesis", during which a complementary strand is built based on the sequence of a template strand.

A Ion Proton semiconductor sequencer.

A microwell containing a template DNA strand to be sequenced is flooded with a single species of deoxyribonucleotide triphosphate (dNTP). If the introduced dNTP is complementary to the leading template nucleotide, it is incorporated into the growing complementary strand. This causes the release of a hydrogen ion that triggers an ISFET ion sensor, which indicates that a reaction has occurred. If homopolymer repeats are present in the template sequence, multiple dNTP molecules will be incorporated in a single cycle. This leads to a corresponding number of released hydrogens and a proportionally higher electronic signal.

This technology differs from other sequencing technologies in that no modified nucleotides or optics are used. Ion semiconductor sequencing may also be referred to as Ion Torrent sequencing, pH-mediated sequencing, silicon sequencing, or semiconductor sequencing.

Technology

Sequencing Chemistry

In nature, the incorporation of a deoxyribonucleoside triphosphate (dNTP) into a growing DNA strand involves the formation of a covalent bond and the release of pyrophosphate and a positively charged hydrogen ion. A dNTP will only be incorporated if it is complementary to the leading unpaired template nucleotide. Ion semiconductor sequencing exploits these facts by determining if a hydrogen ion is released upon providing a single species of dNTP to the reaction.

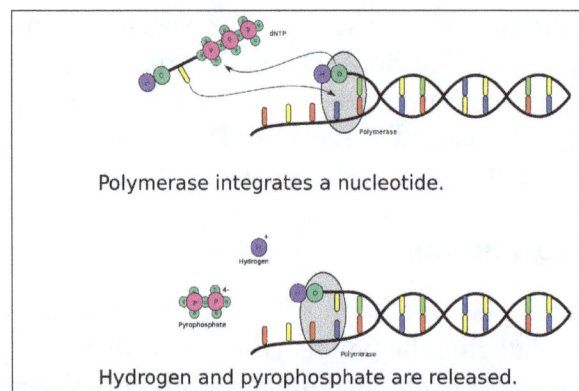

The incorporation of deoxyribonucleotide Triphosphate into a growing DNA strand
causes the release of hydrogen and pyrophosphate.

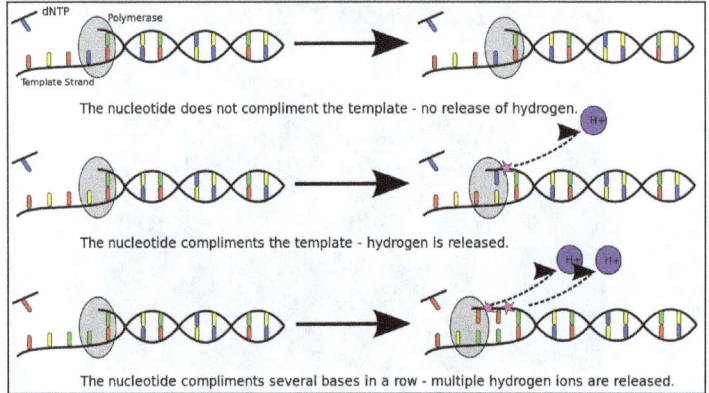

The release of hydrogen ions indicate if zero, one or more nucleotides were incorporated.

Microwells on a semiconductor chip that each contain many copies of one single-stranded template DNA molecule to be sequenced and DNA polymerase are sequentially flooded with unmodified A, C, G or T dNTP. If an introduced dNTP is complementary to the next unpaired nucleotide on the template strand it is incorporated into the growing complementary strand by the DNA polymerase. If the introduced dNTP is not complementary there is no incorporation and no biochemical reaction. The hydrogen ion that is released in the reaction changes the pH of the solution, which is detected by an ISFET. The unattached dNTP molecules are washed out before the next cycle when a different dNTP species is introduced.

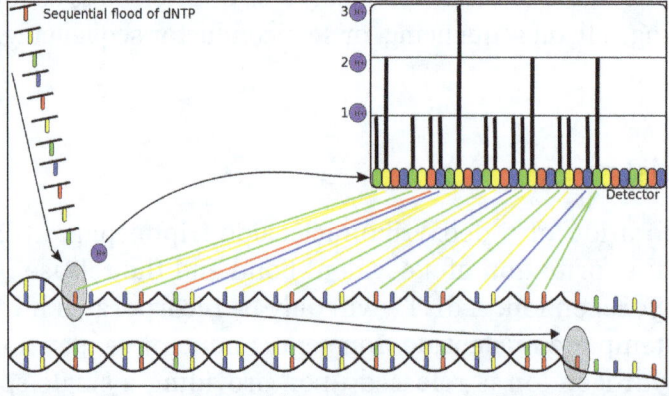

Released hydrogens ions are detected by an ion sensor. Multiple incorporations lead to
a corresponding number of released hydrogens and intensity of signal.

Signal Detection

Beneath the layer of microwells is an ion sensitive layer, below which is an ISFET ion sensor. All layers are contained within a CMOS semiconductor chip, similar to that used in the electronics industry.

Each chip contains an array of microwells with corresponding ISFET detectors. Each released hydrogen ion then triggers the ISFET ion sensor. The series of electrical pulses transmitted from the chip to a computer is translated into a DNA sequence, with no intermediate signal conversion required. Because nucleotide incorporation events are measured directly by electronics, the use of labeled nucleotides and optical measurements are avoided. Signal processing and DNA assembly can then be carried out in software.

Sequencing Characteristics

The per base accuracy achieved in house by Ion Torrent on the Ion Torrent Ion semiconductor sequencer as of February 2011 was 99.6% based on 50 base reads, with 100 Mb per run. The read-length as of February 2011 was 100 base pairs. The accuracy for homopolymer repeats of 5 repeats in length was 98%. Later releases show a read length of 400 base pairs These figures have not yet been independently verified outside of the company.

Strengths

The major benefits of ion semiconductor sequencing are rapid sequencing speed and low upfront and operating costs. This has been enabled by the avoidance of modified nucleotides and optical measurements.

Because the system records natural polymerase-mediated nucleotide incorporation events, sequencing can occur in real-time. In reality, the sequencing rate is limited by the cycling of substrate nucleotides through the system. Ion Torrent Systems Inc., the developer of the technology, claims that each incorporation measurement takes 4 seconds and each run takes about one hour, during which 100-200 nucleotides are sequenced. If the semiconductor chips are improved (as predicted by Moore's law), the number of reads per chip (and therefore per run) should increase.

The cost of acquiring a pH-mediated sequencer from Ion Torrent Systems Inc. at time of launch was priced at around $50,000 USD, excluding sample preparation equipment and a server for data analysis. The cost per run is also significantly lower than that of alternative automated sequencing methods, at roughly $1,000.

Limitations

If homopolymer repeats of the same nucleotide (e.g. TTTTT) are present on the template strand (strand to be sequenced) then multiple introduced nucleotides are incorporated and more hydrogen ions are released in a single cycle. This results in a greater pH change and a proportionally greater electronic signal. This is a limitation of the system in that it is difficult to enumerate long repeats. This limitation is shared by other techniques that detect single nucleotide additions such as pyrosequencing. Signals generated from a high repeat number are difficult to differentiate from repeats of a similar but different number; *e.g.*, homorepeats of length 7 are difficult to differentiate from those of length 8.

Another limitation of this system is the short read length compared to other sequencing methods such as Sanger sequencing or pyrosequencing. Longer read lengths are beneficial for *de novo* genome assembly. Ion Torrent semiconductor sequencers produce an average read length of approximately 400 nucleotides per read.

The throughput is currently lower than that of other high-throughput sequencing technologies, although the developers hope to change this by increasing the density of the chip.

Application

The developers of Ion Torrent semiconductor sequencing have marketed it as a rapid, compact and economical sequencer that can be utilized in a large number of laboratories as a bench top machine. The company hopes that their system will take sequencing outside of specialized centers and into the reach of hospitals and smaller laboratories.

Due to the ability of alternative sequencing methods to achieve a greater read length (and therefore being more suited to whole genome analysis) this technology may be best suited to small scale applications such as microbial genome sequencing, microbial transcriptome sequencing, targeted sequencing, amplicon sequencing, or for quality testing of sequencing libraries.

Nanopore Sequencing

Nanopore sequencing is a third generation approach used in the sequencing of biopolymers- specifically, polynucleotides in the form of DNA or RNA.

Using nanopore sequencing, a single molecule of DNA or RNA can be sequenced without the need for PCR amplification or chemical labeling of the sample. Nanopore sequencing has the potential to offer relatively low-cost genotyping, high mobility for testing, and rapid processing of samples with the ability to display results in real-time. Publications on the method outline its use in rapid identification of viral pathogens, monitoring ebola, environmental monitoring, food safety monitoring, human genome sequencing, plant genome sequencing, monitoring of antibiotic resistance, haplotyping and other applications.

Principles for Detection and Base Identification

Nanopore sequencing uses electrophoresis to transport an unknown sample through an orifice of 10^{-9} meters in diameter. A nanopore system always contains an electrolytic solution, when a constant electric field is applied, an electric current can be observed in the system. The magnitude of the electric current density across a nanopore surface depends on the nanopore's dimensions and the composition of DNA or RNA that is occupying the nanopore. Sequencing is made possible because, when close enough to nanopores, samples cause characteristic changes in electric current density across nanopore surfaces. The total charge flowing through a nanopore channel is equal to the surface integral of electric current density flux across the nanopore unit normal surfaces between times t_1 and t_2.

Types

Biological

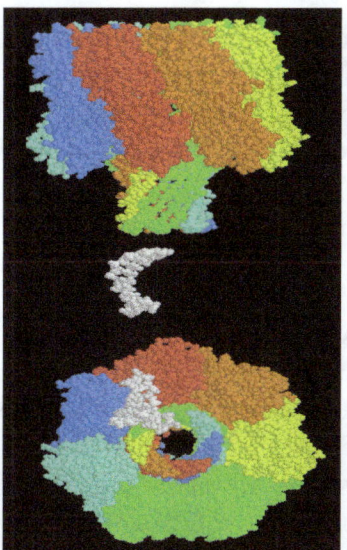

alpha-hemolysin pore (made up of 7 identical subunits in 7 colors) and 12-mer single-stranded DNA (in white) on the same scale to illustrate DNA effects on conductance when moving through a nanopore. Below is an orthogonal view of the same molecules.

Biological nanopore sequencing relies on the use of transmembrane proteins, called porins, that are embedded in lipid membranes so as to create size dependent porous surfaces- with nanometer scale "holes" distributed across the membranes. Sufficiently low translocation velocity can be attained through the incorporation of various proteins that facilitate the movement of DNA or RNA through the pores of the lipid membranes.

Alpha Hemolysin

Alpha hemolysin (αHL), a nanopore from bacteria that causes lysis of red blood cells, has been studied for over 15 years. To this point, studies have shown that all four bases can be identified using ionic current measured across the αHL pore. The structure of αHL is advantageous to identify specific bases moving through the pore. The αHL pore is ~10 nm long, with two distinct 5 nm sections. The upper section consists of a larger, vestibule-like structure and the lower section consists of three possible recognition sites (R1, R2, R3), and is able to discriminate between each base.

Sequencing using αHL has been developed through basic study and structural mutations, moving towards the sequencing of very long reads. Protein mutation of αHL has improved the detection abilities of the pore. The next proposed step is to bind an exonuclease onto the αHL pore. The enzyme would periodically cleave single bases, enabling the pore to identify successive bases. Coupling an exonuclease to the biological pore would slow the translocation of the DNA through the pore, and increase the accuracy of data acquisition.

Notably, theorists have shown that sequencing via exonuclease enzymes as described here is not feasible. This is mainly due to diffusion related effects imposing a limit on the capture probability of each nucleotide as it is cleaved. This results in a significant probability that a nucleotide is either not captured before it diffuses into the bulk or captured out of order, and therefore is not properly sequenced by the nanopore, leading to insertion and deletion errors. Therefore, major changes are needed to this method before it can be considered a viable strategy.

A recent study has pointed to the ability of αHL to detect nucleotides at two separate sites in the lower half of the pore. The R1 and R2 sites enable each base to be monitored twice as it moves through the pore, creating 16 different measurable ionic current values instead of 4. This method improves upon the single read through the nanopore by doubling the sites that the sequence is read per nanopore.

MspA

Mycobacterium smegmatis porin A (MspA) is the second biological nanopore currently being investigated for DNA sequencing. The MspA pore has been identified as a potential improvement over αHL due to a more favorable structure. The pore is described as a goblet with a thick rim and a diameter of 1.2 nm at the bottom of the pore. A natural MspA, while favorable for DNA sequencing because of shape and diameter, has a negative core that prohibited single stranded DNA (ssDNA) translocation. The natural nanopore was modified to improve translocation by replacing three negatively charged aspartic acids with neutral asparagines.

The electric current detection of nucleotides across the membrane has been shown to be tenfold more specific than αHL for identifying bases. Utilizing this improved specificity, a group at the University of Washington has proposed using double stranded DNA (dsDNA) between each single stranded molecule to hold the base in the reading section of the pore. The dsDNA would halt the base in the correct section of the pore and enable identification of the nucleotide. A recent grant has been awarded to a collaboration from UC Santa Cruz, the University of Washington, and Northeastern University to improve the base recognition of MspA using phi29 polymerase in conjunction with the pore.

Solid State

Solid state nanopore sequencing approaches, unlike biological nanopore sequencing, do not incorporate proteins into their systems. Instead, solid state nanopore technology uses various metal or metal alloy substrates with nanometer sized pores that allow DNA or RNA to pass through. These substrates most often serve integral roles in the sequence recognition of nucleic acids as they translocate through the channels along the substrates.

Tunneling Current

Measurement of electron tunneling through bases as ssDNA translocates through the nanopore is an improved solid state nanopore sequencing method. Most research has focused on proving bases could be determined using electron tunneling. These studies were conducted using a scanning probe microscope as the sensing electrode, and have proved that bases can be identified by specific tunneling currents. After the proof of principle research, a functional system must be created to couple the solid state pore and sensing devices.

Researchers at the Harvard Nanopore group have engineered solid state pores with single walled carbon nanotubes across the diameter of the pore. Arrays of pores are created and chemical vapor deposition is used to create nanotubes that grow across the array. Once a nanotube has grown across a pore, the diameter of the pore is adjusted to the desired size. Successful creation of a nanotube coupled with a pore is an important step towards identifying bases as the ssDNA translocates through the solid state pore.

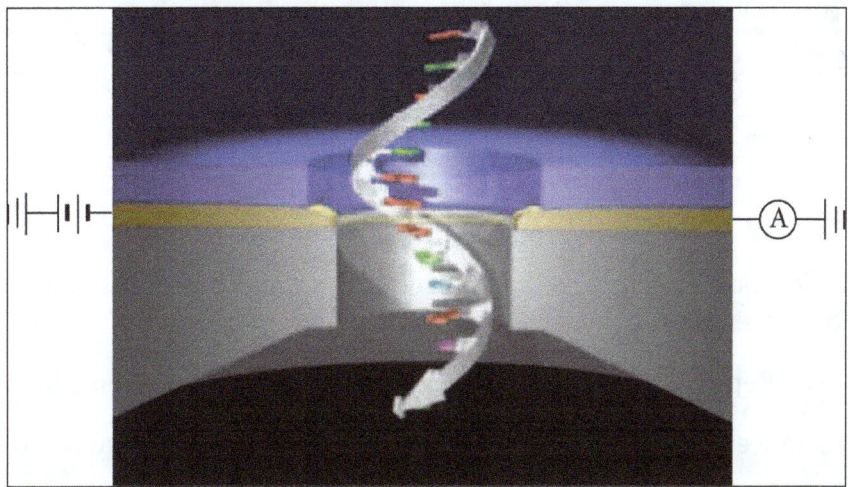

The theoretical movement of ssDNA through a tunneling current nanopore system. Detection is made possible by the incorporation of electrodes along the nanopore channel walls, perpendicular to the ssDNA's velocity vector.

Another method is the use of nanoelectrodes on either side of a pore. The electrodes are specifically created to enable a solid state nanopore's formation between the two electrodes. This technology could be used to not only sense the bases but to help control base translocation speed and orientation.

Fluorescence

An effective technique to determine a DNA sequence has been developed using solid state nanopores and fluorescence. This fluorescence sequencing method converts each base into a characteristic representation of multiple nucleotides which bind to a fluorescent probe strand-forming dsDNA. With the two color system proposed, each base is identified by two separate fluorescences, and will therefore be converted into two specific sequences. Probes consist of a fluorophore and quencher at the start and end of each sequence, respectively. Each fluorophore will be extinguished by the quencher at the end of the preceding sequence. When the dsDNA is translocating through a solid state nanopore, the probe strand will be stripped off, and the upstream fluorophore will fluoresce.

This sequencing method has a capacity of 50-250 bases per second per pore, and a four color fluorophore system (each base could be converted to one sequence instead of two), will sequence over 500 bases per second. Advantages of this method are based on the clear sequencing readout—using a camera instead of noisy current methods. However, the method does require sample preparation to convert each base into an expanded binary code before sequencing. Instead of one base being identified as it translocates through the pore, ~12 bases are required to find the sequence of one base.

Comparison between Types

Comparison of Biological and Solid State Nanopore Sequencing Systems Based on Major Constraints		
	Biological	Solid State
Low Translocation Velocity	✓	
Dimensional Reproducibility	✓	
Stress Tolerance		✓
Longevity		✓
Ease of Fabrication		✓

Major Constraints

1. Low Translocation Velocity: The speed at which a sample passes through a unit's pore slow enough to be measured.

2. Dimensional Reproducibility: The likelihood of a unit's pore to be made the proper size.

3. Stress Tolerance: The sensitivity of a unit to internal environmental conditions.

4. Longevity: The length of time that a unit is expected to remain functioning.

5. Ease of Fabrication: The ability to produce a unit- usually in regards to mass-production.

Advantages and Disadvantages

Biological nanopore sequencing systems have several fundamental characteristics that make them advantageous as compared with solid state systems, with each advantageous characteristic of this design approach stemming from the incorporation of proteins into their technology. Uniform pore structure, the precise control of sample translocation through pore channels, and even the detection of individual nucleotides in samples can be facilitated by unique proteins from a variety of organism types.

The use of proteins in biological nanopore sequencing systems, despite the various benefits, also brings with it some negative characteristics. The sensitivity of the proteins in these systems to local environmental stress has a large impact on the longevity of the units, overall. One example is that a motor protein may only unzip samples with sufficient speed at a certain pH range while not operating fast enough outside of the range- this constraint impacts the functionality of the whole sequencing unit. Another example is that a transmembrane porin may only operate reliably for a certain number of runs before it breaks down. Both of these examples would have to be controlled for in the design of any viable biological nanopore system- something that may be difficult to achieve while keeping the costs of such a technology as low and as competitive, to other systems, as possible.

Challenges

One challenge for the 'strand sequencing' method was in refining the method to improve its resolution to be able to detect single bases. In the early papers methods, a nucleotide needed to be repeated in a sequence about 100 times successively in order to produce a measurable characteristic change. This low resolution is because the DNA strand moves rapidly at the rate of 1 to 5 μs per base through the nanopore. This makes recording difficult and prone to background noise, failing in obtaining single-nucleotide resolution. The problem is being tackled by either improving the recording technology or by controlling the speed of DNA strand by various protein engineering strategies and Oxford Nanopore employs a 'kmer approach', analyzing more than one base at any one time so that stretches of DNA are subject to repeat interrogation as the strand moves through the nanopore one base at a time. Various techniques including algorithmic have been used to improve the performance of the MinION technology since it was first made available to users. More recently effects of single bases due to secondary structure or released mononucleotides have been shown.

Professor Hagan Bayley proposed in 2010 that creating two recognition sites within an alpha hemolysin pore may confer advantages in base recognition.

One challenge for the 'exonuclease approach', where a processive enzyme feeds individual bases, in the correct order, into the nanopore, is to integrate the exonuclease and the nanopore detection systems. In particular, the problem is that when an exonuclease hydrolyzes the phosphodiester bonds between nucleotides in DNA, the subsequently released nucleotide is not necessarily guaranteed to directly move into, say, a nearby alpha-hemolysin nanopore. One idea is to attach the exonuclease to the nanopore, perhaps through biotinylation to the beta barrel hemolysin. The central pore of the protein may be lined with charged residues arranged so that the positive and negative charges appear on opposite sides of the pore. However, this mechanism is primarily discriminatory and does not constitute a mechanism to guide nucleotides down some particular path.

Pyrosequencing

Pyrosequencing is a method of DNA sequencing (determining the order of nucleotides in DNA) based on the "sequencing by synthesis" principle, in which the sequencing is performed by detecting the nucleotide incorporated by a DNA polymerase. Pyrosequencing relies on light detection based on a chain reaction when pyrophosphate is released. Hence, the name pyrosequencing.

The principle of Pyrosequencing was first described in 1993 by Bertil Pettersson, Mathias Uhlen and Pål Nyren by combining the solid phase sequencing method using streptavidin coated magnetic beads with recombinant DNA polymerase lacking 3´ to 5´ exonuclease activity (proof-reading) and luminescence detection using the firefly luciferase enzyme. A mixture of three enzymes (DNA polymerase, ATP sulfurylase and firefly luciferase) and a nucleotide (dNTP) are added to single stranded DNA to be sequenced and the incorporation of nucleotide is followed by measuring the light emitted. The intensity of the light determines if 0, 1 or more nucleotides have been incorporated, thus showing how many complementary nucleotides are present on the template strand. The nucleotide mixture is removed before the next nucleotide mixture is added. This process is repeated with each of the four nucleotides until the DNA sequence of the single stranded template is determined.

A second solution-based method for Pyrosequencing was described in 1998 by Mostafa Ronaghi, Mathias Uhlen and Pål Nyren. In this alternative method, an additional enzyme apyrase is introduced to remove nucleotides that are not incorporated by the DNA polymerase. This enabled the enzyme mixture including the DNA polymerase, the luciferase and the apyrase to be added at the start and kept throughout the procedure, thus providing a simple set-up suitable for automation. An automated instrument based on this principle was introduced to the market the following year by the company Pyrosequencing.

A third microfluidic variant of the Pyrosequencing method was described in 2005 by Jonathan Rothberg and co-workers. This alternative approach for Pyrosequencing was based on the original principle of attaching the DNA to be sequenced to a solid support and they showed that sequencing could be performed in a highly parallel manner using a microfabricated microarray. This allowed for high-throughput DNA sequencing and an automated instrument was introduced to the market. This became the first next generation sequencing instrument starting a new era in genomics research, with rapidly falling prices for DNA sequencing allowing whole genome sequencing at affordable prices.

Procedure

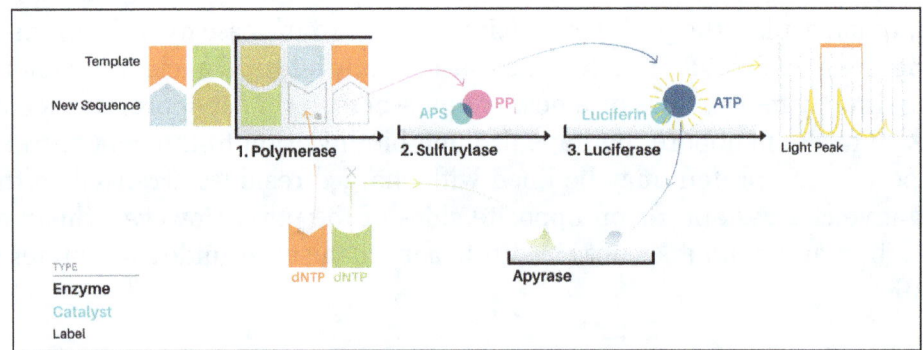

The chart shows how pyrosequencing works.

"Sequencing by synthesis" involves taking a single strand of the DNA to be sequenced and then synthesizing its complementary strand enzymatically. The pyrosequencing method is based on detecting the activity of DNA polymerase (a DNA synthesizing enzyme) with another chemoluminescent enzyme. Essentially, the method allows sequencing a single strand of DNA by synthesizing the complementary strand along it, one base pair at a time, and detecting which base was actually added at each step. The template DNA is immobile, and solutions of A, C, G, and T nucleotides are sequentially added and removed from the reaction. Light is produced only when the nucleotide solution complements the first unpaired base of the template. The sequence of solutions which produce chemiluminescent signals allows the determination of the sequence of the template.

For the solution-based version of Pyrosequencing, the single-strand DNA (ssDNA) template is hybridized to a sequencing primer and incubated with the enzymes DNA polymerase, ATP sulfurylase, luciferase and apyrase, and with the substrates adenosine 5´ phosphosulfate (APS) and luciferin.

1. The addition of one of the four deoxynucleotide triphosphates (dNTPs) (dATPαS, which is not a substrate for a luciferase, is added instead of dATP to avoid noise) initiates the second step. DNA polymerase incorporates the correct, complementary dNTPs onto the template. This incorporation releases pyrophosphate (PPi).

2. ATP sulfurylase converts PPi to ATP in the presence of adenosine 5´ phosphosulfate. This ATP acts as a substrate for the luciferase-mediated conversion of luciferin to oxyluciferin that generates visible light in amounts that are proportional to the amount of ATP. The light produced in the luciferase-catalyzed reaction is detected by a camera and analyzed in a program.

3. Unincorporated nucleotides and ATP are degraded by the apyrase, and the reaction can restart with another nucleotide.

Limitations

Currently, a limitation of the method is that the lengths of individual reads of DNA sequence are in the neighborhood of 300-500 nucleotides, shorter than the 800-1000 obtainable with chain termination methods (e.g. Sanger sequencing). This can make the process of genome assembly more difficult, particularly for sequences containing a large amount of repetitive DNA. Lack of proof-reading activity limits accuracy of this method.

Single-molecule Real-time Sequencing

Single-molecule real-time sequencing (SMRT) is a parallelized single molecule DNA sequencing method. Single-molecule real-time sequencing utilizes a zero-mode waveguide (ZMW). A single DNA polymerase enzyme is affixed at the bottom of a ZMW with a single molecule of DNA as a template. The ZMW is a structure that creates an illuminated observation volume that is small enough to observe only a single nucleotide of DNA being incorporated by DNA polymerase. Each of the four DNA bases is attached to one of four different fluorescent dyes. When a nucleotide is incorporated by the DNA polymerase, the fluorescent tag is cleaved off and diffuses out of the observation area of the ZMW where its fluorescence is no longer observable. A detector detects the fluorescent signal of the nucleotide incorporation, and the base call is made according to the corresponding fluorescence of the dye.

Technology

The DNA sequencing is done on a chip that contains many ZMWs. Inside each ZMW, a single active DNA polymerase with a single molecule of single stranded DNA template is immobilized to the bottom through which light can penetrate and create a visualization chamber that allows monitoring of the activity of the DNA polymerase at a single molecule level. The signal from a phospho-linked nucleotide incorporated by the DNA polymerase is detected as the DNA synthesis proceeds which results in the DNA sequencing in real time.

Phospholinked Nucleotide

For each of the nucleotide bases, there is a corresponding fluorescent dye molecule that enables the detector to identify the base being incorporated by the DNA polymerase as it performs the DNA synthesis. The fluorescent dye molecule is attached to the phosphate chain of the nucleotide. When the nucleotide is incorporated by the DNA polymerase, the fluorescent dye is cleaved off with the phosphate chain as a part of a natural DNA synthesis process during which a phosphodiester bond is created to elongate the DNA chain. The cleaved fluorescent dye molecule then diffuses out of the detection volume so that the fluorescent signal is no longer detected.

Zero-mode Waveguide

The zero-mode waveguide (ZMW) is a nanophotonic confinement structure that consists of a circular hole in an aluminum cladding film deposited on a clear silica substrate.

The ZMW holes are ~70 nm in diameter and ~100 nm in depth. Due to the behavior of light when it travels through a small aperture, the optical field decays exponentially inside the chamber.

The observation volume within an illuminated ZMW is ~20 zeptoliters (20×10^{-21} liters). Within this volume, the activity of DNA polymerase incorporating a single nucleotide can be readily detected.

Sequencing Performance

Sequencing performance for the technology can be measured in read length and total throughput per experiment.

On 19 Sep 2018, Pacific Biosciences [PacBio] released the Sequel 6.0 chemistry, synchronizing the chemistry version with the software version. Performance is contrasted for large-insert libraries with high molecular weight DNA versus shorter-insert libraries below ~15,000 bases in length. For larger templates average read lengths are up to 30,000 bases. For shorter-insert libraries, average read length are up to 100,000 bases while reading the same molecule in a circle. The latter shorter-insert libraries then yield up to 50 billion bases from a single SMRT Cell.

Application

Single-molecule real-time sequencing may be applicable for a broad range of genomics research.

For *de novo* genome sequencing, read lengths from the single-molecule real-time sequencing are comparable to or greater than that from the Sanger sequencing method based on dideoxynucleotide chain termination. The longer read length allows *de novo* genome sequencing and easier genome assemblies. Scientists are also using single-molecule real-time sequencing in hybrid assemblies for de novo genomes to combine short-read sequence data with long-read sequence data. In 2012, several peer-reviewed publications were released demonstrating the automated finishing of bacterial genomes, including one paper that updated the Celera Assembler with a pipeline for genome finishing using long SMRT sequencing reads. In 2013, scientists estimated that long-read sequencing could be used to fully assemble and finish the majority of bacterial and archaeal genomes.

The same DNA molecule can be resequenced independently by creating the circular DNA template and utilizing a strand displacing enzyme that separates the newly synthesized DNA strand from the template. In August 2012, scientists from the Broad Institute published an evaluation of SMRT sequencing for SNP calling.

The dynamics of polymerase can indicate whether a base is methylated. Scientists demonstrated the use of single-molecule real-time sequencing for detecting methylation and other base modifications. In 2012 a team of scientists used SMRT sequencing to generate the full methylomes of six bacteria. In November 2012, scientists published a report on genome-wide methylation of an outbreak strain of E. coli.

Long reads make it possible to sequence full gene isoforms, including the 5' and 3' ends. This type of sequencing is useful to capture isoforms and splice variants.

Single Cell Sequencing

Single cell sequencing examines the sequence information from individual cells with optimized next generation sequencing (NGS) technologies, providing a higher resolution of cellular differences and a better understanding of the function of an individual cell in the context of its microenvironment. Sequencing the DNA of individual cells can give information about mutations carried by small populations of cells, for example in cancer, while sequencing the RNAs expressed by individual cells can give insight into the existence and behavior of different cell types, for example in development.

A typical human cell consists of about 2×3.3 billion base pairs of DNA and 600 million bases of mRNA. Usually a mix of millions of cells are used in sequencing the DNA or RNA using traditional methods like Sanger sequencing or Illumina sequencing. By using deep sequencing of DNA and RNA from a single cell, cellular functions can be investigated extensively. Like typical NGS experiments, the protocols of single cell sequencing generally contain the following steps: isolation of a single cell, nucleic acid extraction and amplification, sequencing library preparation, sequencing and bioinformatic data analysis. It is more challenging to perform single cell sequencing in comparison with sequencing from cells in bulk. The minimal amount of starting materials from a single cell make degradation, sample loss and contamination exert pronounced effects on quality of sequencing data. In addition, due to the picogram level of the amount of nucleic acids used, heavy amplification is often needed during sample preparation of single cell sequencing, resulting in the uneven coverage, noise and inaccurate quantification of sequencing data.

Recent technical improvements make single cell sequencing a promising tool for approaching a set of seemingly inaccessible problems. For example, heterogeneous samples, rare cell types, cell lineage relationships, mosaicism of somatic tissues, analyses of microbes that cannot be cultured, and disease evolution can all be elucidated through single cell sequencing.

Single Cell Genome Sequencing

Single cell DNA genome sequencing involves isolating a single cell, performing whole-genome-amplification (WGA), constructing sequencing libraries and then sequencing the DNA using a next-generation sequencer (ex. Illumina, Ion Torrent). A genome constructed in this fashion is commonly referred to as a single amplified genome or SAG. It can be used in microbiome studies, in order to obtain genomic data from uncultured microorganisms. In addition, it can be united with high throughput cell sorting of microorganisms and cancer. One popular method used for single cell genome sequencing is multiple displacement amplification and this enables research into various areas such as microbial genetics, ecology and infectious diseases. Furthermore, data obtained from microorganisms might establish processes for culturing in the future. Some of the genome assembly tools that can be used in single cell genome sequencing include: SPAdes, IDBA-UD, Cortex and HyDA.

Method

Multiple displacement amplification (MDA) is a widely used technique, enabling amplifying femtograms of DNA from bacterium to micrograms for the use of sequencing. Reagents required for MDA reactions include: random primers and DNA polymerase from bacteriophage phi29. In 30 degree isothermal reaction, DNA is amplified with included reagents. As the polymerases manufacture new strands, a strand displacement reaction takes place, synthesizing multiple copies from each template DNA. At the same time, the strands that were extended antecedently will be displaced. MDA products result in a length of about 12 kb and ranges up to around 100 kb, enabling its use in DNA sequencing. In 2017, a major improvement to this technique, called WGA-X, was introduced by taking advantage of a thermostable mutant of the phi29 polymerase, leading to better genome recovery from individual cells, in particular those with high G+C content. Other methods include MALBAC.

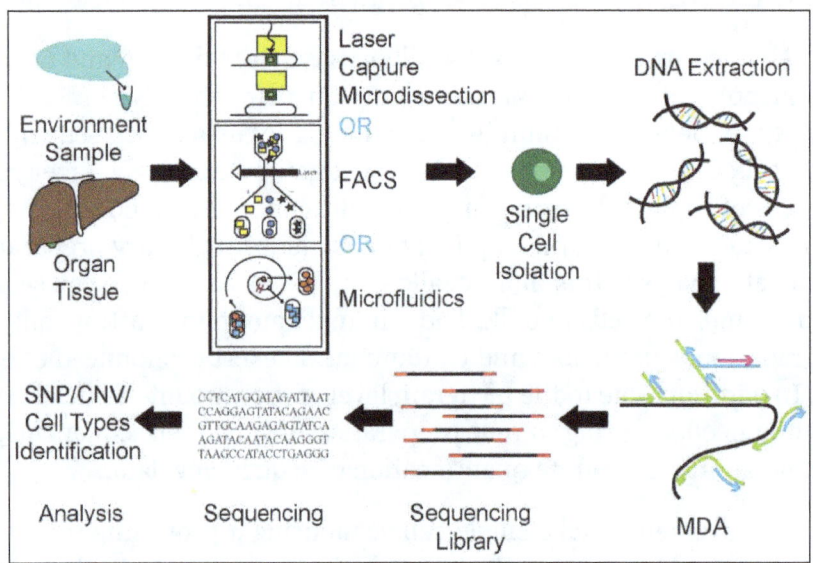

The steps involved in workflow of single cell genome sequencing.
MDA stands for multiple displacement amplification.

Limitations

MDA of individual cell genomes results in highly uneven genome coverage, i.e. relative overrepresentation and underrepresentation of various regions of the template, leading to loss of some sequences. There are two components to this process: a) stochastic over- and under-amplification of random regions; and b) systematic bias against high %GC regions. The stochastic component may be addressed by pooling single-cell MDA reactions from the same cell type, by employing fluorescent in situ hybridization (FISH) and post-sequencing confirmation. The bias of MDA against high %GC regions can be addressed by using thermostable polymerases, such as in the process called WGA-X.

Single-nucleotide polymorphisms (SNPs), which are a big part of genetic variation in the human genome, and copy number variation (CNV), pose problems in single cell sequencing, as well as the limited amount of DNA extracted from a single cell. Due to scant amounts of DNA, accurate analysis of DNA poses problems even after amplification since coverage is low and susceptible to errors. With MDA, average genome coverage is less than 80% and SNPs that are not covered by sequencing reads will be opted out. In addition, MDA shows a high ratio of allele dropout, not detecting alleles from heterozygous samples. Various SNP algorithms are currently in use but none are specific to single cell sequencing. MDA with CNV also poses the problem of identifying false CNVs that conceal the real CNVs. To solve this, when patterns can be generated from false CNVs, algorithms can detect and eradicate this noise to produce true variants.

Applications

Microbiomes are among the main targets of single cell genomics due to the difficulty of culturing the majority of microorganisms in most environments. Single cell genomics is a powerful way to obtain microbial genome sequences without cultivation. This approach has been widely applied on marine, soil, subsurface, organismal, and other types of microbiomes in order to address a wide array of questions related to microbial ecology, evolution, public health and biotechnology potential.

Cancer sequencing is also an emerging application of scDNAseq. Fresh or frozen tumors may be analyzed and categorized with respect to SCNAs, SNVs, and rearrangements quite well using whole genome DNAS approaches. Cancer scDNAseq is particularly useful for examining the depth of complexity and compound mutations present in amplified therapeutic targets such as receptor tyrosine kinase genes (EGFR, PDGFRA etc.) where conventional population-level approaches of the bulk tumor are not able to resolve the co-occurrence patterns of these mutations within single cells of the tumor. Such overlap may provide redundancy of pathway activation and tumor cell resistance.

Single Cell DNA Methylome Sequencing

Single cell DNA methylome sequencing quantifies DNA methylation. This is similar to single cell genome sequencing, but with the addition of a bisulfite treatment before sequencing. Forms include whole genome bisulfite sequencing, and reduced representation bisulfite sequencing.

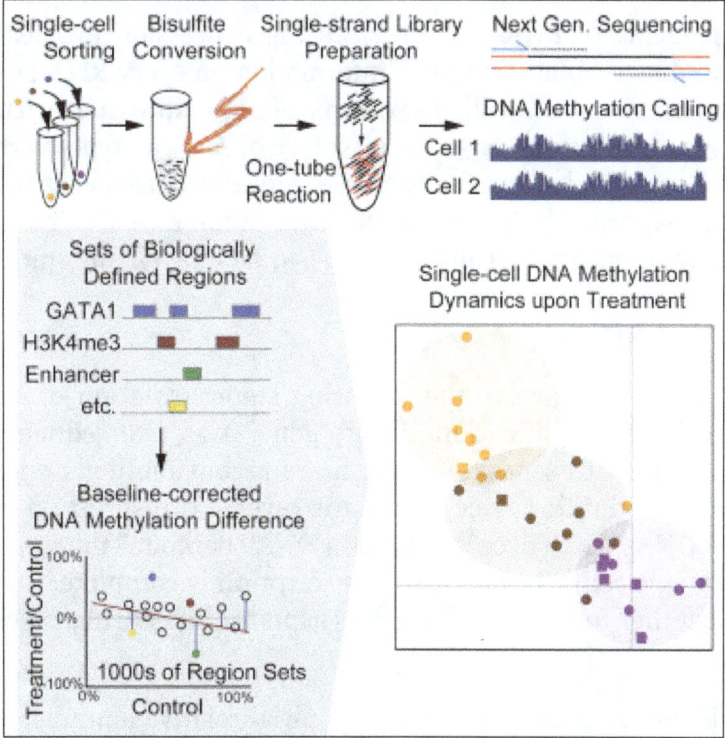

One method for single cell DNA methylation sequencing.

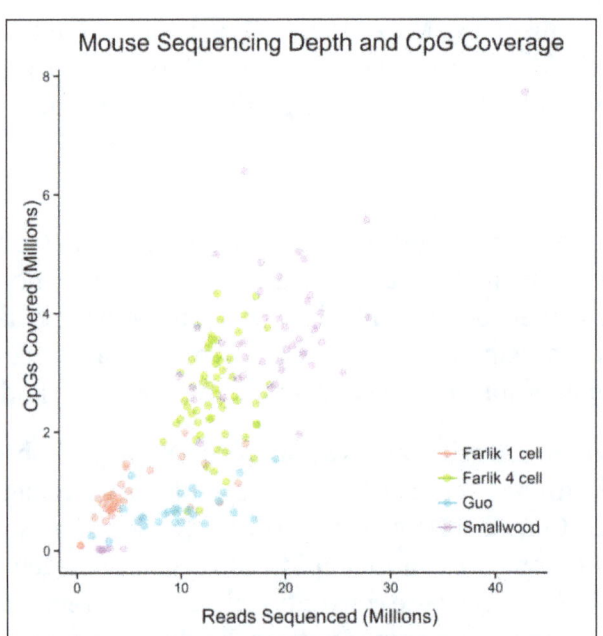

Comparison of single cell methylation sequencing methods in terms
of coverage as at 2015 on Mus musculus.

Single Cell RNA Sequencing

Standard methods such as microarrays and standard bulk RNA-seq analysis analyze the expression of RNAs from large populations of cells. In mixed cell populations, these measurements may obscure critical differences between individual cells within these populations.

Single-cell RNA sequencing (scRNA-seq) provides the expression profiles of individual cells. Although it is not possible to obtain complete information on every RNA expressed by each cell, due to the small amount of material available, patterns of gene expression can be identified through gene clustering analyses. This can uncover the existence of rare cell types within a cell population that may never have been seen before. For example, rare specialized cells in the lung called pulmonary ionocytes that express the Cystic Fibrosis Transmembrane Conductance Regulator were identified in 2018 by two groups performing scRNA-Seq on lung airway epithelia.

Experimental Procedures

Current scRNA-seq protocols involve the following steps: isolation of single cell and RNA, reverse transcription (RT), amplification, library generation and sequencing. Early methods separated individual cells into separate wells; more recent methods encapsulate individual cells in droplets in a microfluidic device, where the reverse transcription reaction takes place, converting RNAs to cDNAs. Each droplet carries a DNA "barcode" that uniquely labels the cDNAs derived from a single cell. Once reverse transcription is complete, the cDNAs from many cells can be mixed together for sequencing; transcripts from a particular cell are identified by the unique barcode.

Challenges for scRNA-Seq include preserving the initial relative abundance of mRNA in a cell and identifying rare transcripts. The reverse transcription step is critical as the efficiency of the

RT reaction determines how much of the cell's RNA population will be eventually analyzed by the sequencer. The processivity of reverse transcriptases and the priming strategies used may affect full-length cDNA production and the generation of libraries biased toward 3' or 5' end of genes.

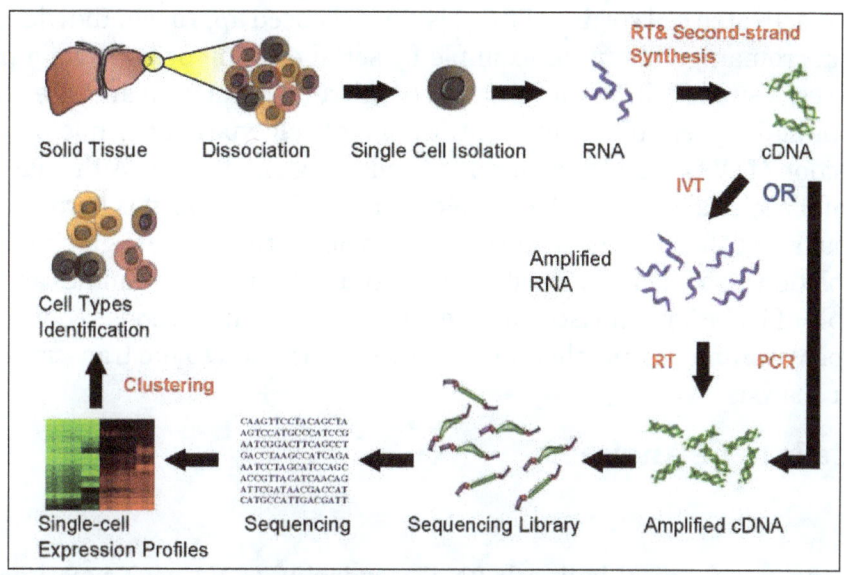

Single-cell RNA sequencing workflow.

In the amplification step, either PCR or in vitro transcription (IVT) is currently used to amplify cDNA. One of the advantages of PCR-based methods is the ability to generate full-length cDNA. However, different PCR efficiency on particular sequences (for instance, GC content and snapback structure) may also be exponentially amplified, producing libraries with uneven coverage. On the other hand, while libraries generated by IVT can avoid PCR-induced sequence bias, specific sequences may be transcribed inefficiently, thus causing sequence dropout or generating incomplete sequences. Several scRNA-seq protocols have been published: Tang et al., STRT, SMART-seq, CEL-seq, RAGE-seq, Quartz-seq. and C1-CAGE. These protocols differ in terms of strategies for reverse transcription, cDNA synthesis and amplification, and the possibility to accommodate sequence-specific barcodes (i.e. UMIs) or the ability to process pooled samples.

In 2017, two approaches were introduced to simultaneously measure single-cell mRNA and protein expression through oligonucleotide-labeled antibodies known as REAP-seq, and CITE-seq.

Applications

scRNA-Seq is becoming widely used across biological disciplines including Development, Neurology, Oncology, Autoimmune disease, and Infectious disease.

scRNA-Seq has provided considerable insight into the development of embryos and organisms, including the worm Caenorhabditis elegans, and the regenerative planarian Schmidtea mediterranea. The first vertebrate animals to be mapped in this way were Zebrafish and *Xenopus laevis*. In each case multiple stages of the embryo were studied, allowing the entire process of development to be mapped on a cell-by-cell basis.

Considerations

Isolation of Single Cells

There are several ways to isolate individual cells prior to whole genome amplification and sequencing. Fluorescence-activated cell sorting (FACS) is a widely used approach. Individual cells can also be collected by micromanipulation, for example by serial dilution or by using a patch pipette or nanotube to harvest a single cell. The advantages of micromanipulation are ease and low cost, but they are laborious and susceptible to misidentification of cell types under microscope. Laser-capture microdissection (LCM) can also be used for collecting single cells. Although LCM preserves the knowledge of the spatial location of a sampled cell within a tissue, it is hard to capture a whole single cell without also collecting the materials from neighboring cells. High-throughput methods for single cell isolation also include microfluidics. Both FACS and microfluidics are accurate, automatic and capable of isolating unbiased samples. However, both methods require detaching cells from their microenvironments first, thereby causing perturbation to the transcriptional profiles in RNA expression analysis.

Number of Cells to be Analyzed

scRNA-Seq

Generally speaking, for a typical bulk cell RNA-sequencing (RNA-seq) experiment, ten million reads are generated and a gene with higher than the threshold of 50 reads per kb per million reads (RPKM) is considered expressed. For a gene that is 1kb long, this corresponds to 500 reads and a minimum coefficient of variation (CV) of 4% under the assumption of the Poisson distribution. For a typical mammalian cell containing 200,000 mRNA, sequencing data from at least 50 single cells need to be pooled in order to achieve this minimum CV value. However, due to the efficiency of reverse transcription and other noise introduced in the experiments, more cells are required for accurate expression analyses and cell type identification.

Single Cell Transcriptomics

Single cell transcriptomics examines the gene expression level of individual cells in a given population by simultaneously measuring the messenger RNA (mRNA) concentration of hundreds to thousands of genes. The unraveling of heterogenous cell populations, reconstruction of cellular developmental trajectories, and modeling of transcriptional dynamics — all previously masked in bulk transcriptome measurements — are made possible through analysis of this transcriptomic data.

Gene expression analysis has become routine through the development of high-throughput RNA sequencing (RNA-seq) and microarrays. RNA analysis that was previously limited to tracing individual transcripts by Northern blots or quantitative PCR is now used frequently to characterize the expression profiles of populations of thousands of cells.The data produced from the bulk based assays has led to the identification of genes that are differentially expressed in distinct cell populations and biomarker discovery.

These genomic studies are limited as they provide measurements for whole tissues and as a result show an average expression profile for all the constituent cells. In multicellular organisms different cell types within the same population can have distinct roles and form subpopulations with different transcriptional profiles. Correlations in the gene expression of the subpopulations can often be missed due to the lack of subpopulation identification. Moreover, bulk assays fail to identify if a change in the expression profile is due to a change in regulation or composition, in which one cell type arises to dominate the population. Lastly, when examining cellular progression through differentiation, average expression profiles are only able to order cells by time rather than their stage of development and are consequently unable to show trends in gene expression levels specific to certain stages.

Recent advances in biotechnology allow the measurement of gene expression in hundreds to thousands of individual cells simultaneously. Whilst these breakthroughs in transcriptomics technologies have enabled the generation of single-cell transcriptomic data there are new computational and analytical challenges presented by the data produced. Techniques used for analysing RNA-seq data from bulk cell populations can be used for single-cell data but many new computational approaches have been designed for this data type to facilitate a complete and detailed study of single-cell expression profiles.

Experimental Steps

There is currently no standardized technique to generate single-cell data, all methods must include cell isolation from the population, lysate formation, amplification through reverse transcription and quantification of expression levels. Common techniques for measuring expression are quantitative PCR or RNA-seq.

Isolating Single Cells

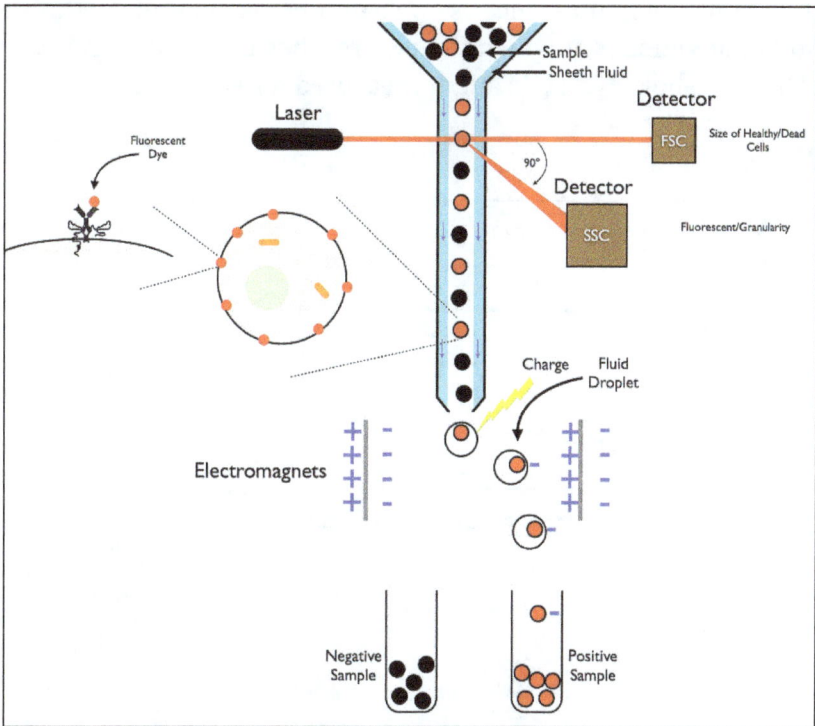

Fluorescence Assisted Cell Sorting workflow (FACS).

There are several methods available to isolate and amplify cells for single-cell analysis. Low throughput techniques are able to isolate hundreds of cells, are slow and enable selection. These methods include:

- Micropipetting.

- Cytoplasmic aspiration.

- Laser capture microdissection.

High throughput methods are able to quickly isolate hundreds to tens of thousands of cells. Common techniques include:

- Fluorescence activated cell sorting (FACS).

- Microfluidic devices.

Quantitative PCR

To measure the level of expression of each transcript qPCR can be applied. Gene specific primers are used to amplify the corresponding gene as with regular PCR and as a result data is usually only obtained for sample sizes of less than 100 genes. The inclusion of housekeeping genes, whose expression should be constant under the conditions, is used for normalisation. The most commonly used house keeping genes include GAPDH and α-actin, although the reliability of normalisation through this process is questionable as there is evidence that the level of expression can vary significantly. Fluorescent dyes are used as reporter molecules to detect the PCR product and monitor the progress of the amplification - the increase in fluorescence intensity is proportional to the amplicon concentration. A plot of fluorescence vs. cycle number is made and a threshold fluorescence level is used to find cycle number at which the plot reaches this value. The cycle number at this point is known as the threshold cycle (C_t) and is measured for each gene.

Single Cell RNA-seq

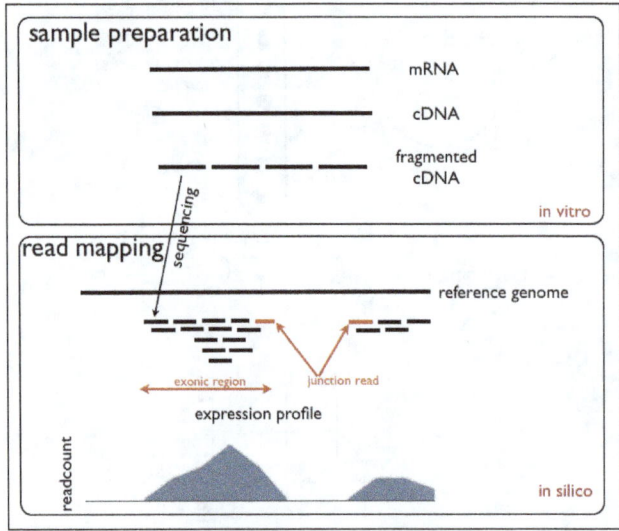

RNA Seq Experiment.

The Single cell RNA-seq technique converts a population of RNAs to a library of cDNA fragments. These fragments are sequenced by high-throughput next generation sequencing techniques and the reads are mapped back to the reference genome, providing a count of the number of reads associated with each gene.

Normalisation of RNA-seq data accounts for cell to cell variation in the efficiencies of the cDNA library formation and sequencing. One method relies on the use of extrinsic RNA spike-ins (RNA sequences of known sequence and quantity) that are added in equal quantities to each cell lysate and used to normalise read count by the number of reads mapped to spike-in mRNA.

Another control uses unique molecular identifiers (UMIs)-short DNA sequences (6–10nt) that are added to each cDNA before amplification and act as a bar code for each cDNA molecule. Normalisation is achieved by using the count number of unique UMIs associated with each gene to account for differences in amplification efficiency.

A combination of both spike-ins, UMIs and other approaches have been combined for more accurate normalisation.

Considerations

A problem associated with single-cell data occurs in the form of zero inflated gene expression distributions, known as technical dropouts, that are common due to low mRNA concentrations of less-expressed genes that are not captured in the reverse transcription process. The percentage of mRNA molecules in the cell lysate that are detected is often only 10-20%.

When using RNA spike-ins for normalisation the assumption is made that the amplification and sequencing efficiencies for the endogenous and spike-in RNA are the same. Evidence suggests that this is not the case given fundamental differences in size and features, such as the lack of a polyadenylated tail in spike-ins and therefore shorter length. Additionally, normalisation using UMIs assumes the cDNA library is sequenced to saturation, which is not always the case.

Data Analysis

Insights based on single-cell data analysis assumes that the input is a matrix of normalised gene expression counts, generated by the approaches outline above, and can provide opportunities that are not obtainable by bulk.

Three main insights provided:

1. Identification and characterization of cell types and their spatial organisation in time.

2. Inference of gene regulatory networks and their strength across individual cells.

3. Classification of the stochastic component of transcription.

The techniques outlined have been designed to help visualise and explore patterns in the data in order to facilitate the revelation of these three features.

Clustering

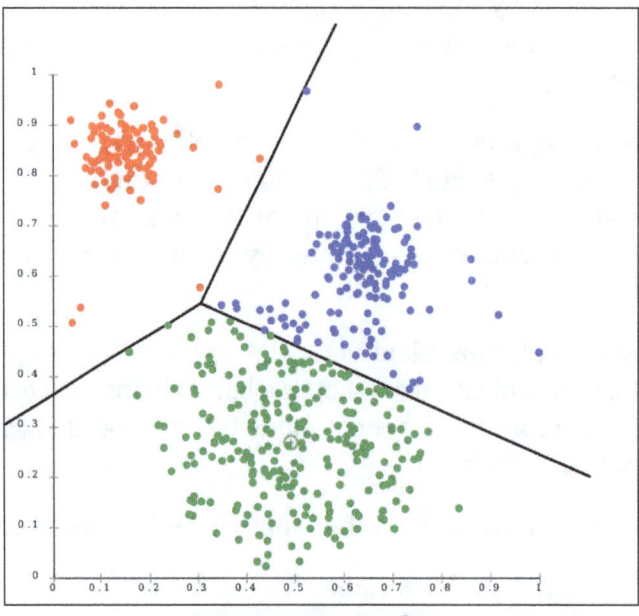

K-means-Gaussian-data.

Clustering allows for the formation of subgroups in the cell population. Cells can be clustered by their transcriptomic profile in order to analyse the sub-population structure and identify rare cell types or cell subtypes. Alternatively, genes can be clustered by their expression states in order to identify covarying genes. A combination of both clustering approaches, known as biclustering, has been used to simultaneously cluster by genes and cells to find genes that behave similarly within cell clusters.

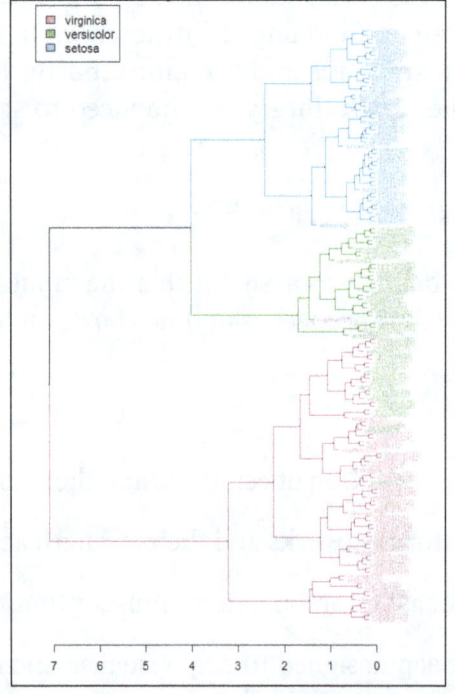

Iris dendrogram produced using a Hierarchical clustering algorithm.

Clustering methods applied can be K-means clustering, forming disjoint groups or hierarchical clustering, forming nested partitions.

Biclustering

Biclustering provides several advantages by improving the resolution of clustering. Genes that are only informative to a subset of cells and are hence only expressed there can be identified through biclustering. Moreover, similarly behaving genes that differentiate one cell cluster from another can be identified using this method.

Dimensionality Reduction

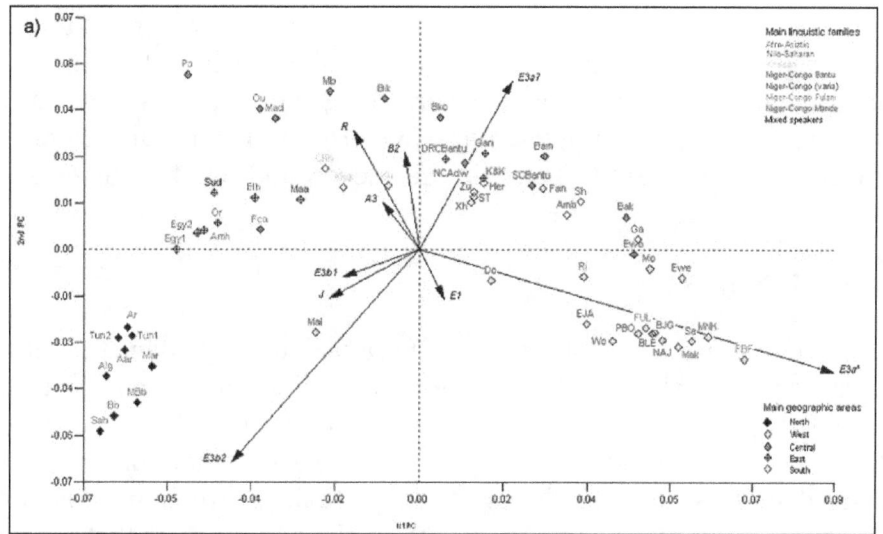

PCA example of Guinean and other African populations Y chromosome haplogroup frequencies.

Dimensionality reduction algorithms such as Principal component analysis (PCA) and t-SNE can be used to simplify data for visualisation and pattern detection by transforming cells from a high to a lower dimensional space. The result of this method produces graphs with each cell as a point in a 2-D or 3-D space. Dimensionality reduction is frequently used before clustering as cells in high dimensions can wrongly appear to be close due to distance metrics behaving non-intuitively.

Principal Component Analysis

The most frequently used technique is PCA, which identifies the directions of largest variance principal components and transforms the data so that the first principal component has the largest possible variance, and successive principle components in turn each have the highest variance possible while remaining orthogonal to the preceding components. The contribution each gene makes to each component is used to infer which genes are contributing the most to variance in the population and are involved in differentiating different subpopulations.

Differential Expression

Detecting differences in gene expression level between two populations is used both single-cell and bulk transcriptomic data. Specialised methods have been designed for single-cell data that considers single cell features such as technical dropouts and shape of the distribution e.g. Bimodal vs. unimodal.

Gene Ontology Enrichment

Gene ontology terms describe gene functions and the relationships between those functions into three classes:

1. Molecular function.

2. Cellular component.

3. Biological process.

Gene Ontology (GO) term enrichment is a technique used to identify which GO terms are over-represented or under-represented in a given set of genes. In single-cell analysis input list of genes of interest can be selected based on differentially expressed genes or groups of genes generated from biclustering. The number of genes annotated to a GO term in the input list is normalised against the number of genes annotated to a GO term in the background set of all genes in genome to determine statistical significance.

Pseudotemporal Ordering

Pseudo-temporal ordering (or trajectory inference) is a technique that aims to infer gene expression dynamics from snapshot single-cell data. The method tries to order the cells in such a way that similar cells are closely positioned to each other. This trajectory of cells can be linear, but can also bifurcate or follow more complex graph structures. The trajectory therefore enables the inference of gene expression dynamics and the ordering of cells by their progression through differentiation or response to external stimuli. The method relies on the assumptions that the cells follow the same path through the process of interest and that their transcriptional state correlates to their progression. The algorithm can be applied to both mixed populations and temporal samples.

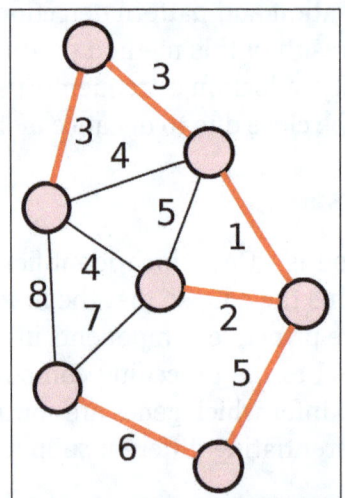

Graph with minimal spanning tree.

More than 50 methods for pseudo-temporal ordering have been developed, and each has its own requirements for prior information (such as starting cells or time course data), detectable topologies, and methodology. An example algorithm is the Monocle algorithm that carries out dimensionality reduction of the data, builds a minimal spanning tree using the transformed data, orders cells in pseudo-time by following the longest connected path of the tree and consequently labels cells by type. Another example is DPT, which uses a diffusion map and diffusion process to infer the pseudotime on a bifurcating trajectory.

Network Inference

Gene regulatory network inference is a technique that aims to construct a network, shown as a graph, in which the nodes represent the genes and edges indicate co-regulatory interactions. The method relies on the assumption that a strong statistical relationship between the expression of genes is an indication of a potential functional relationship. The most commonly used method to measure the strength of a statistical relationship is correlation. However, correlation fails to identify non-linear relationships and mutual information is used as an alternative. Gene clusters linked in a network signify genes that undergo coordinated changes in expression.

MOLECULAR CLONING

Molecular cloning refers to the replication and recombination of DNA molecules. The first cloning experiments were carried out in the 1970s, when restriction endonucleases were discovered. Restriction endonucleases are enzymes that selectively cleave and cut DNA like a pair of scissors.

These enzymes allowed scientists to cut out a specific piece of DNA, usually a gene, and paste it into a vector where it could be copied. In the process, changes can also be made to the cloned DNA segment. Cloning became a foundational molecular biology method, and is the basis for much of what is known today about genetics.

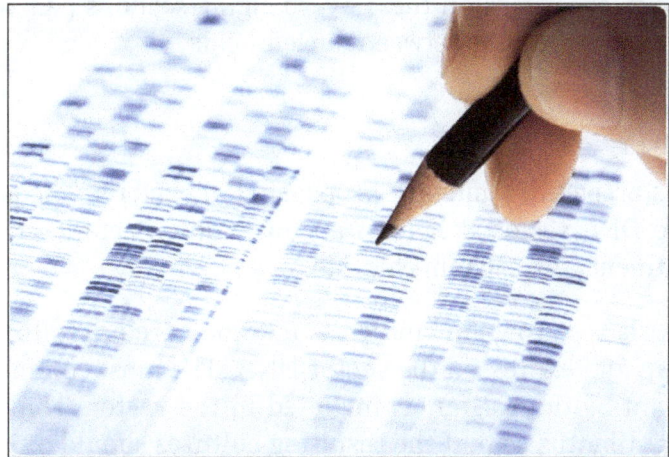

Scientist analyzes DNA gel used in genetics, forensics, drug discovery, biology and medicine.

The basic process of cloning is the isolation of a DNA sequence from any species and insertion into a piece of carrier DNA, called a vector, for propagation inside a host species, which is usually bacteria or yeast. Those clones can then be used for genetic analysis, to produce protein, for mutation, and many other purposes.

The workflow of cloning has four steps:

1. DNA is isolated.

2. Isolated DNA is inserted into a cloning vector.

3. Vectors are propagated in a host.

4. Hosts containing the DNA insert are selected.

Isolation of DNA

A cloning experiment begins by identifying the section of DNA that is being targeted for cloning. Often, this is a complete gene. The most common method to copy the gene is by polymerase chain reaction (PCR). PCR amplifies only the desired section of DNA.

Once the DNA is amplified, the ends are prepared for ligation with restriction enzymes. Restriction enzymes are chosen for their compatibility with the desired vector. The amplified gene segment is then incubated with the restriction enzymes to cut the ends so there is an appropriate overhang for ligation.

At the same time, a vector is prepared to receive the DNA with the same overhanging ends.

The vector usually contains other genes that are useful for cloning, such as promoters and genes for antibiotic resistance that can be used for selection.

Ligation

During ligation, the amplified gene to be cloned is incubated with the vector. The ends of each have been prepared with restriction enzymes to have complementary overhanging pieces of single-stranded DNA. The nucleic acids on these overhanging sections pair with each other, and a ligase enzyme is used to close the gaps between nucleic acids.

Transformation

Once the DNA vector is prepared, it must be propagated in bacteria. The process of inducing the bacteria to take up the DNA vector is called transformation. Bacteria are usually prepared for transformation by treatment with calcium chloride.

The transformed bacteria are grown in culture. A method is required to select only those bacteria that have taken up and expressed the vector DNA. The most common method is through the use of a gene for antibiotic resistance, included in the vector. The growth media for the bacteria contains the antibiotic, so that the resulting cultures should only contain successfully transformed bacteria.

The first molecular cloning experiment was completed in 1973. The method has now become commonplace, and many innovations have been introduced to make molecular cloning faster, easier, more reliable, or customizable for specific types of genes and applications.

DNA CLONING

DNA cloning is the process of making multiple, identical copies of a particular piece of DNA. In a typical DNA cloning procedure, the gene or other DNA fragment of interest (perhaps a gene for a medically important human protein) is first inserted into a circular piece of DNA called a plasmid. The insertion is done using enzymes that "cut and paste" DNA, and it produces a molecule of re-combinant DNA, or DNA assembled out of fragments from multiple sources.

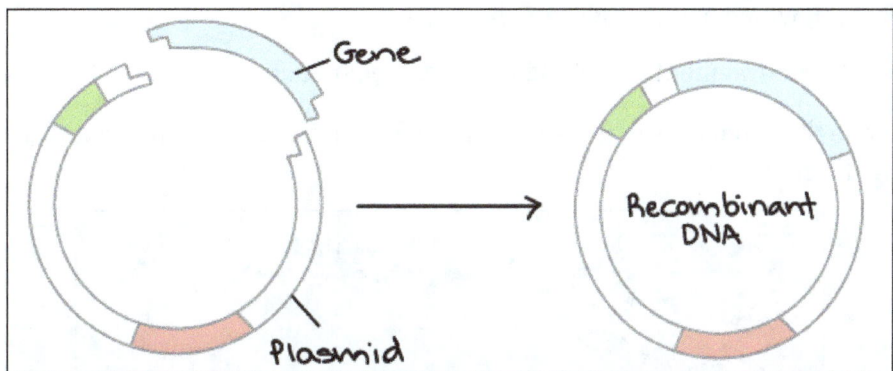

Next, the recombinant plasmid is introduced into bacteria. Bacteria carrying the plasmid are selected and grown up. As they reproduce, they replicate the plasmid and pass it on to their offspring, making copies of the DNA it contains.

What is the point of making many copies of a DNA sequence in a plasmid? In some cases, we need lots of DNA copies to conduct experiments or build new plasmids. In other cases, the piece of DNA encodes a useful protein, and the bacteria are used as "factories" to make the protein. For instance, the human insulin gene is expressed in E. coli bacteria to make insulin used by diabetics.

Steps of DNA Cloning

DNA cloning is used for many purposes. As an example, let's see how DNA cloning can be used to synthesize a protein (such as human insulin) in bacteria. The basic steps are:

1. Cut open the plasmid and "paste" in the gene. This process relies on restriction enzymes (which cut DNA) and DNA ligase (which joins DNA).

2. Insert the plasmid into bacteria. Use antibiotic selection to identify the bacteria that took up the plasmid.

3. Grow up lots of plasmid-carrying bacteria and use them as "factories" to make the protein. Harvest the protein from the bacteria and purify it.

Let's take a closer look at each step.

Cutting and Pasting DNA

How can pieces of DNA from different sources be joined together? A common method uses two types of enzymes: restriction enzymes and DNA ligase.

A restriction enzyme is a DNA-cutting enzyme that recognizes a specific target sequence and cuts DNA into two pieces at or near that site. Many restriction enzymes produce cut ends with short, single-stranded overhangs. If two molecules have matching overhangs, they can base-pair and stick together. However, they won't combine to form an unbroken DNA molecule until they are joined by DNA ligase, which seals gaps in the DNA backbone.

Our goal in cloning is to insert a target gene (e.g., for human insulin) into a plasmid. Using a carefully chosen restriction enzyme, we digest:

- The plasmid, which has a single cut site.

- The target gene fragment, which has a cut site near each end.

Then, we combine the fragments with DNA ligase, which links them to make a recombinant plasmid containing the gene.

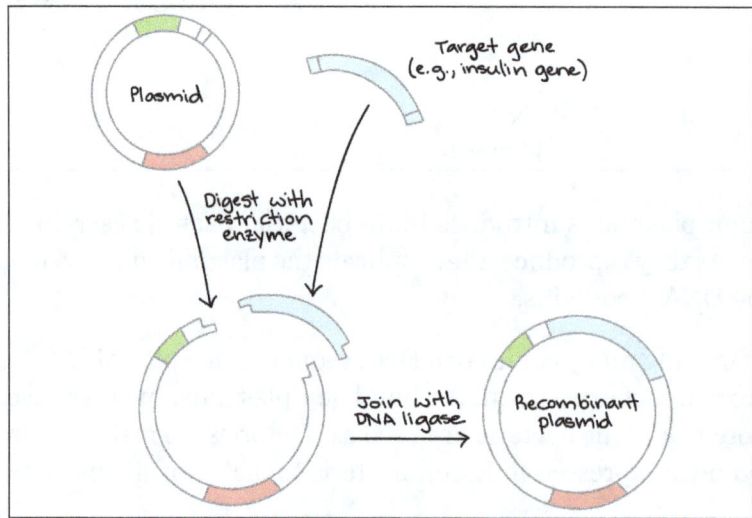

Bacterial Transformation and Selection

Plasmids and other DNA can be introduced into bacteria, such as the harmless E. coli used in labs, in a process called transformation. During transformation, specially prepared bacterial cells are given a shock (such as high temperature) that encourages them to take up foreign DNA.

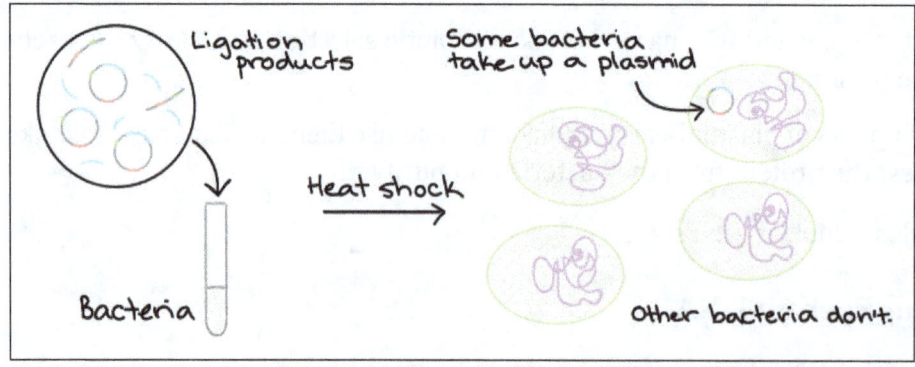

A plasmid typically contains an antibiotic resistance gene, which allows bacteria to survive in the presence of a specific antibiotic. Thus, bacteria that took up the plasmid can be selected on nutrient plates containing the antibiotic. Bacteria without a plasmid will die, while bacteria carrying a plasmid can live and reproduce. Each surviving bacterium will give rise to a small, dot-like group, or colony, of identical bacteria that all carry the same plasmid.

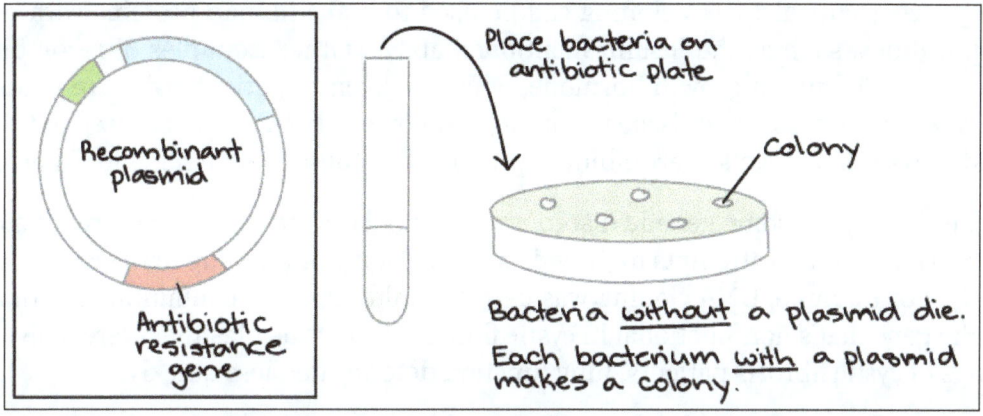

Not all colonies will necessarily contain the right plasmid. That's because, during a ligation, DNA fragments don't always get "pasted" in exactly the way we intend. Instead, we must collect DNA from several colonies and see whether each one contain the right plasmid. Methods like restriction enzyme digestion and PCR are commonly used to check the plasmids.

Protein Production

Once we have found a bacterial colony with the right plasmid, we can grow a large culture of plasmid-bearing bacteria. Then, we give the bacteria a chemical signal that instructs them to make the target protein.

The bacteria serve as miniature "factories," churning out large amounts of protein. For instance, if our plasmid contained the human insulin gene, the bacteria would start transcribing the gene and translating the mRNA to produce many molecules of human insulin protein.

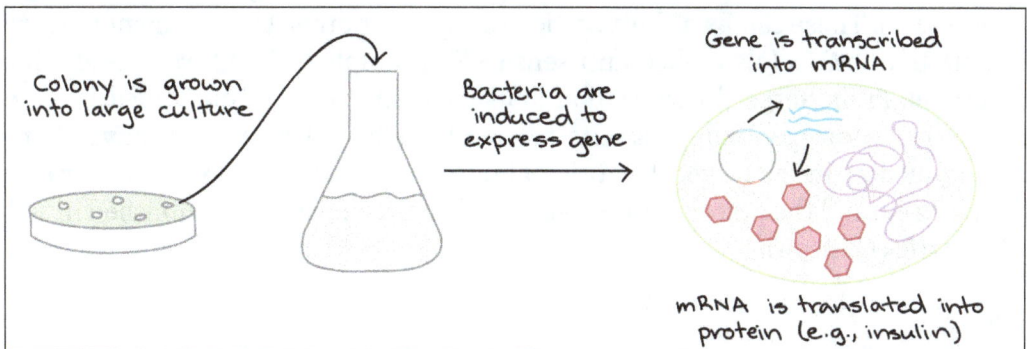

Once the protein has been produced, the bacterial cells can be split open to release it. There are many other proteins and macromolecules floating around in bacteria besides the target protein (e.g., insulin). Because of this, the target protein must be purified, or separated from the other contents of the cells by biochemical techniques. The purified protein can be used for experiments or, in the case of insulin, administered to patients.

Uses of DNA Cloning

DNA molecules built through cloning techniques are used for many purposes in molecular biology. A short list of examples includes:

- Biopharmaceuticals: DNA cloning can be used to make human proteins with biomedical applications, such as the insulin mentioned above. Other examples of recombinant proteins include human growth hormone, which is given to patients who are unable to synthesize the hormone, and tissue plasminogen activator (tPA), which is used to treat strokes and prevent blood clots. Recombinant proteins like these are often made in bacteria.

- Gene therapy: In some genetic disorders, patients lack the functional form of a particular gene. Gene therapy attempts to provide a normal copy of the gene to the cells of a patient's body. For example, DNA cloning was used to build plasmids containing a normal version of the gene that's nonfunctional in cystic fibrosis. When the plasmids were delivered to the lungs of cystic fibrosis patients, lung function deteriorated less quickly.

- Gene analysis: In basic research labs, biologists often use DNA cloning to build artificial, recombinant versions of genes that help them understand how normal genes in an organism function.

These are just a few examples of how DNA cloning is used in biology today. DNA cloning is a very common technique that is used in a huge variety of molecular biology applications.

DNA MICROARRAY

A DNA microarray (also commonly known as DNA chip or biochip) is a collection of microscopic DNA spots attached to a solid surface. Scientists use DNA microarrays to measure the expression levels of large numbers of genes simultaneously or to genotype multiple regions of a genome. Each DNA spot contains picomoles (10^{-12} moles) of a specific DNA sequence, known as *probes* (or *reporters* or *oligos*). These can be a short section of a gene or other DNA element that are used to hybridize a cDNA or cRNA (also called anti-sense RNA) sample (called *target*) under high-stringency conditions. Probe-target hybridization is usually detected and quantified by detection of fluorophore-, silver-, or chemiluminescence-labeled targets to determine relative abundance of nucleic acid sequences in the target. The original nucleic acid arrays were macro arrays approximately 9 cm × 12 cm and the first computerized image based analysis was published in 1981. It was invented by Patrick O. Brown.

Principle

The core principle behind microarrays is hybridization between two DNA strands, the property of complementary nucleic acid sequences to specifically pair with each other by forming hydrogen bonds between complementary nucleotide base pairs. A high number of complementary base pairs in a nucleotide sequence means tighter non-covalent bonding between the two strands. After washing off non-specific bonding sequences, only strongly paired strands will remain hybridized.

Fluorescently labeled target sequences that bind to a probe sequence generate a signal that depends on the hybridization conditions (such as temperature), and washing after hybridization. Total strength of the signal, from a spot (feature), depends upon the amount of target sample binding to the probes present on that spot. Microarrays use relative quantitation in which the intensity of a feature is compared to the intensity of the same feature under a different condition, and the identity of the feature is known by its position.

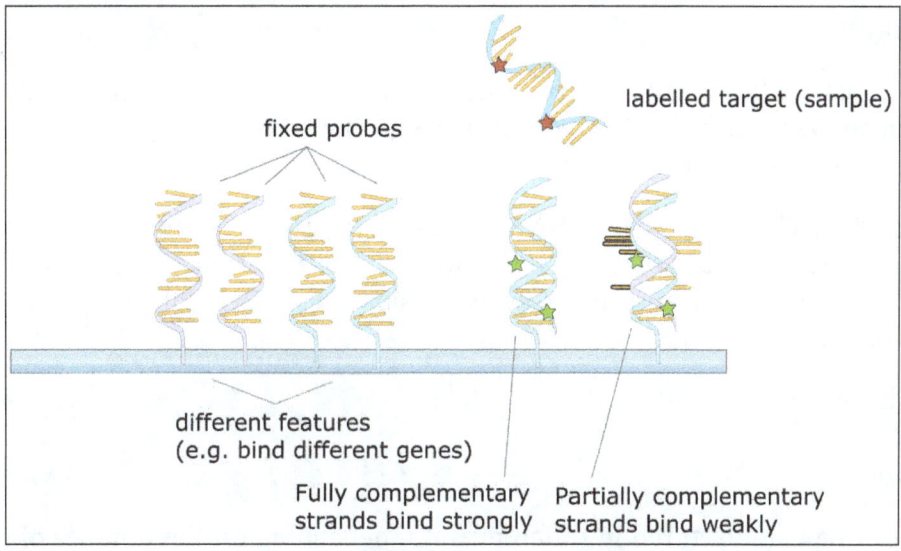

Hybridization of the target to the probe.

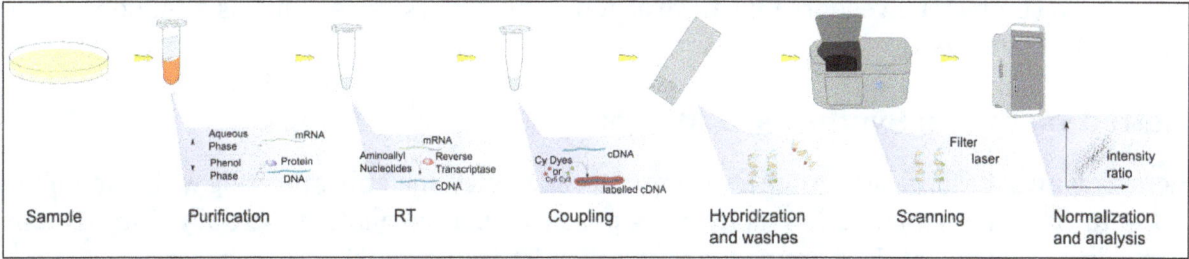

The steps required in a microarray experiment.

Uses and Types

Two Affymetrix chips. A match is shown at bottom left for size comparison.

Many types of arrays exist and the broadest distinction is whether they are spatially arranged on a surface or on coded beads:

- The traditional solid-phase array is a collection of orderly microscopic "spots", called features, each with thousands of identical and specific probes attached to a solid surface, such as glass, plastic or silicon biochip (commonly known as a genome chip, DNA chip or gene array). Thousands of these features can be placed in known locations on a single DNA microarray.

- The alternative bead array is a collection of microscopic polystyrene beads, each with a specific probe and a ratio of two or more dyes, which do not interfere with the fluorescent dyes used on the target sequence.

DNA microarrays can be used to detect DNA (as in comparative genomic hybridization), or detect RNA (most commonly as cDNA after reverse transcription) that may or may not be translated into proteins. The process of measuring gene expression via cDNA is called expression analysis or expression profiling.

Applications include:

Fabrication

Microarrays can be manufactured in different ways, depending on the number of probes under examination, costs, customization requirements, and the type of scientific question being asked. Arrays from commercial vendors may have as few as 10 probes or as many as 5 million or more micrometre-scale probes.

Spotted vs. In Situ Synthesised Arrays

Microarrays can be fabricated using a variety of technologies, including printing with fine-pointed pins onto glass slides, photolithography using pre-made masks, photolithography using dynamic micromirror devices, ink-jet printing, or electrochemistry on microelectrode arrays.

A DNA microarray being printed by a robot at the University of Delaware.

In *spotted microarrays*, the probes are oligonucleotides, cDNA or small fragments of PCR products that correspond to mRNAs. The probes are synthesized prior to deposition on the array surface and are then "spotted" onto glass. A common approach utilizes an array of fine pins or needles controlled by a robotic arm that is dipped into wells containing DNA probes and then depositing each probe at designated locations on the array surface. The resulting "grid" of probes represents the nucleic acid profiles of the prepared probes and is ready to receive complementary cDNA or cRNA "targets" derived from experimental or clinical samples. This technique is used by research scientists around the world to produce "in-house" printed microarrays from their own labs. These arrays may be easily customized for each experiment, because researchers can choose the probes and printing locations on the arrays, synthesize the probes in their own lab (or collaborating facility), and spot the arrays. They can then generate their own labeled samples for hybridization, hybridize the samples to the array, and finally scan the arrays with their own equipment. This provides a relatively low-cost microarray that may be customized for each study, and avoids the costs of purchasing often more expensive commercial arrays that may represent vast numbers of genes that are not of interest to the investigator. Publications exist which indicate in-house spotted microarrays may not provide the same level of sensitivity compared to commercial oligonucleotide arrays, possibly owing to the small batch sizes and reduced printing efficiencies when compared to industrial manufactures of oligo arrays.

In *oligonucleotide microarrays*, the probes are short sequences designed to match parts of the sequence of known or predicted open reading frames. Although oligonucleotide probes are often used in "spotted" microarrays, the term "oligonucleotide array" most often refers to a specific technique of manufacturing. Oligonucleotide arrays are produced by printing short oligonucleotide sequences designed to represent a single gene or family of gene splice-variants by synthesizing this sequence directly onto the array surface instead of depositing intact sequences. Sequences may be longer or shorter depending on the desired purpose; longer probes are more specific to individual target genes, shorter probes may be spotted in higher density across the array and are cheaper to manufacture. One technique used to produce oligonucleotide arrays include photolithographic synthesis on a silica substrate where light and light-sensitive masking agents are used to "build" a sequence one nucleotide at a time across the entire array. Each applicable probe is selectively "unmasked" prior to bathing the array in a solution of a single nucleotide, then a masking reaction takes place and the next set of probes are unmasked in preparation for a different nucleotide exposure. After many repetitions, the sequences of every probe become fully constructed. More recently, Maskless Array Synthesis from NimbleGen Systems has combined flexibility with large numbers of probes.

Two-channel vs. One-channel Detection

Two-color microarrays or two-channel microarrays are typically hybridized with cDNA prepared from two samples to be compared (e.g. diseased tissue versus healthy tissue) and that are labeled with two different fluorophores. Fluorescent dyes commonly used for cDNA labeling include Cy3, which has a fluorescence emission wavelength of 570 nm (corresponding to the green part of the light spectrum), and Cy5 with a fluorescence emission wavelength of 670 nm (corresponding to the red part of the light spectrum). The two Cy-labeled cDNA samples are mixed and hybridized to a single microarray that is then scanned in a microarray scanner to visualize fluorescence of the two fluorophores after excitation with a laser beam of a defined wavelength. Relative intensities of each fluorophore may then be used in ratio-based analysis to identify up-regulated and down-regulated genes.

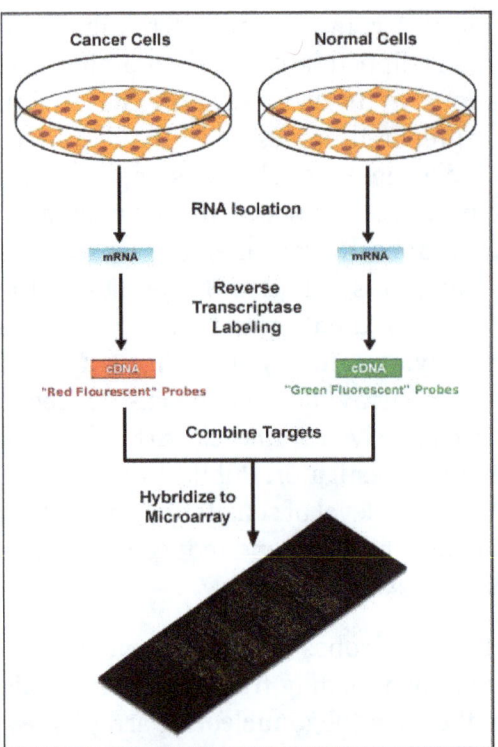

Diagram of typical dual-colour microarray experiment.

Oligonucleotide microarrays often carry control probes designed to hybridize with RNA spike-ins. The degree of hybridization between the spike-ins and the control probes is used to normalize the hybridization measurements for the target probes. Although absolute levels of gene expression may be determined in the two-color array in rare instances, the relative differences in expression among different spots within a sample and between samples is the preferred method of data analysis for the two-color system. Examples of providers for such microarrays includes Agilent with their Dual-mode platform, Eppendorf with their Dual Chip platform for colorimetric Silverquant labeling, and TeleChem International with Arrayit.

In *single-channel microarrays* or *one-color microarrays*, the arrays provide intensity data for each probe or probe set indicating a relative level of hybridization with the labeled target. However, they do not truly indicate abundance levels of a gene but rather relative abundance when compared to other samples or conditions when processed in the same experiment. Each RNA molecule encounters protocol and batch-specific bias during amplification, labeling, and hybridization phases of the experiment making comparisons between genes for the same microarray uninformative. The comparison of two conditions for the same gene requires two separate single-dye hybridizations. Several popular single-channel systems are the Affymetrix "Gene Chip", Illumina "Bead Chip", Agilent single-channel arrays, the Applied Microarrays "CodeLink" arrays, and the Eppendorf "DualChip & Silverquant". One strength of the single-dye system lies in the fact that an aberrant sample cannot affect the raw data derived from other samples, because each array chip is exposed to only one sample (as opposed to a two-color system in which a single low-quality sample may drastically impinge on overall data precision even if the other sample was of high quality). Another benefit is that data are more easily compared to arrays from different experiments as long as batch effects have been accounted for.

One channel microarray may be the only choice in some situations. Suppose samples need to be compared: then the number of experiments required using the two channel arrays quickly becomes unfeasible, unless a sample is used as a reference.

Number of samples	One-channel microarray	Two channel microarray	Two channel microarray
1	1	1	1
2	2	1	1
3	3	3	2
4	4	6	3
i	i	$i(i-1)/2$	$i-1$

A Typical Protocol

This is an example of a DNA microarray experiment, detailing a particular case to better explain DNA microarray experiments, while enumerating possible alternatives.

1. The two samples to be compared (pairwise comparison) are grown/acquired. In this example treated sample (case) and untreated sample (control).

2. The nucleic acid of interest is purified: this can be all RNA for expression profiling, DNA for comparative hybridization, or DNA/RNA bound to a particular protein which is immunoprecipitated (ChIP-on-chip) for epigenetic or regulation studies. In this example total RNA is isolated (total as it is nuclear and cytoplasmic) by Guanidinium thiocyanate-phenol-chloroform extraction (e.g. Trizol) which isolates most RNA (whereas column methods have a cut off of 200 nucleotides) and if done correctly has a better purity.

3. The purified RNA is analysed for quality (by capillary electrophoresis) and quantity (for example, by using a NanoDrop or NanoPhotometer spectrometer). If the material is of acceptable quality and sufficient quantity is present (e.g., > 1 µg, although the required amount varies by microarray platform), the experiment can proceed.

4. The labelled product is generated via reverse transcription and sometimes with an optional PCR amplification. The RNA is reverse transcribed with either polyT primers (which amplify only mRNA) or random primers (which amplify all RNA, most of which is rRNA); miRNA microarrays ligate an oligonucleotide to the purified small RNA (isolated with a fractionator), which is then reverse transcribed and amplified. The label is added either during the reverse transcription step, or following amplification if it is performed. The sense labelling is dependent on the microarray; e.g. if the label is added with the RT mix, the cDNA is antisense and the microarray probe is sense, except in the case of negative controls. The label is typically fluorescent; only one machine uses radiolabels. The labelling can be direct (not used) or indirect (requires a coupling stage). For two-channel arrays, the coupling stage occurs before hybridization, using aminoallyl uridine triphosphate (aminoallyl-UTP, or aaUTP) and NHS amino-reactive dyes (such as cyanine dyes); for single-channel arrays, the coupling stage occurs after hybridization, using biotin and labelled streptavidin. The modified nucleotides (usually in a ratio of 1 aaUTP: 4 TTP (thymidine triphosphate)) are added enzymatically in a low ratio to normal nucleotides, typically resulting in 1 every 60 bases. The aaDNA is then purified with a column (using a phosphate buffer solution, as Tris

contains amine groups). The aminoallyl group is an amine group on a long linker attached to the nucleobase, which reacts with a reactive dye. A form of replicate known as a dye flip can be performed to remove any dye effects in two-channel experiments; for a dye flip, a second slide is used, with the labels swapped (the sample that was labeled with Cy3 in the first slide is labeled with Cy5, and vice versa). In this example, aminoallyl-UTP is present in the reverse-transcribed mixture.

5. The labeled samples are then mixed with a propriety hybridization solution which can consist of SDS, SSC, dextran sulfate, a blocking agent (such as COT1 DNA, salmon sperm DNA, calf thymus DNA, PolyA or PolyT), Denhardt's solution, or formamine.

6. The mixture is denatured and added to the pinholes of the microarray. The holes are sealed and the microarray hybridized, either in a hyb oven, where the microarray is mixed by rotation, or in a mixer, where the microarray is mixed by alternating pressure at the pinholes.

7. After an overnight hybridization, all nonspecific binding is washed off (SDS and SSC).

8. The microarray is dried and scanned by a machine that uses a laser to excite the dye and measures the emission levels with a detector.

9. The image is gridded with a template and the intensities of each feature (composed of several pixels) is quantified.

10. The raw data is normalized; the simplest normalization method is to subtract background intensity and scale so that the total intensities of the features of the two channels are equal, or to use the intensity of a reference gene to calculate the t-value for all of the intensities. More sophisticated methods include z-ratio, loess and lowess regression and RMA (robust multichip analysis) for Affymetrix chips (single-channel, silicon chip, in situ synthesised short oligonucleotides).

Microarrays and Bioinformatics

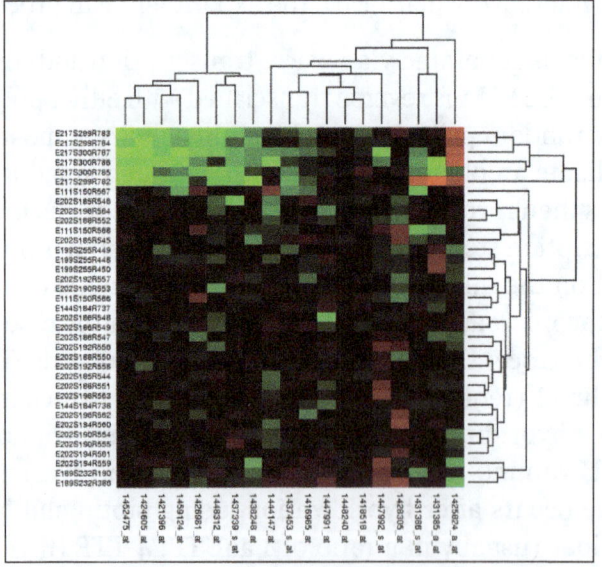

Gene expression values from microarray experiments can be represented
as heat maps to visualize the result of data analysis.

The advent of inexpensive microarray experiments created several specific bioinformatics challenges: the multiple levels of replication in experimental design (Experimental design); the number of platforms and independent groups and data format (Standardization); the statistical treatment of the data (Data analysis); mapping each probe to the mRNA transcript that it measures (Annotation); the sheer volume of data and the ability to share it (Data warehousing).

Experimental Design

Due to the biological complexity of gene expression, the considerations of experimental design are of critical importance if statistically and biologically valid conclusions are to be drawn from the data.

There are three main elements to consider when designing a microarray experiment. First, replication of the biological samples is essential for drawing conclusions from the experiment. Second, technical replicates (two RNA samples obtained from each experimental unit) help to ensure precision and allow for testing differences within treatment groups. The biological replicates include independent RNA extractions and technical replicates may be two aliquots of the same extraction. Third, spots of each cDNA clone or oligonucleotide are present as replicates (at least duplicates) on the microarray slide, to provide a measure of technical precision in each hybridization.

Standardization

Microarray data is difficult to exchange due to the lack of standardization in platform fabrication, assay protocols, and analysis methods. This presents an interoperability problem in bioinformatics. Various grass-roots open-source projects are trying to ease the exchange and analysis of data produced with non-proprietary chips.

For example, the "Minimum Information About a Microarray Experiment" (MIAME) checklist helps define the level of detail that should exist and is being adopted by many journals as a requirement for the submission of papers incorporating microarray results. But MIAME does not describe the format for the information, so while many formats can support the MIAME requirements, as of 2007 no format permits verification of complete semantic compliance. The "MicroArray Quality Control (MAQC) Project" is being conducted by the US Food and Drug Administration (FDA) to develop standards and quality control metrics which will eventually allow the use of MicroArray data in drug discovery, clinical practice and regulatory decision-making. The MGED Society has developed standards for the representation of gene expression experiment results and relevant annotations.

Data Analysis

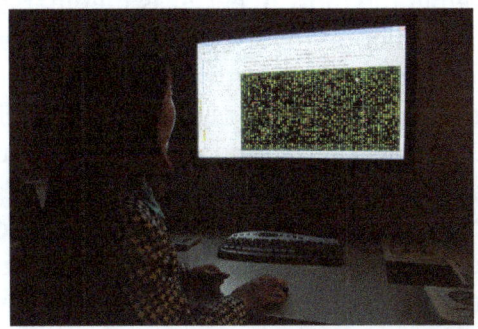

National Center for Toxicological Research scientist reviews microarray data.

Microarray data sets are commonly very large, and analytical precision is influenced by a number of variables. Statistical challenges include taking into account effects of background noise and appropriate normalization of the data. Normalization methods may be suited to specific platforms and, in the case of commercial platforms, the analysis may be proprietary. Algorithms that affect statistical analysis include:

- Image analysis: gridding, spot recognition of the scanned image (segmentation algorithm), removal or marking of poor-quality and low-intensity features (called flagging).

- Data processing: background subtraction (based on global or local background), determination of spot intensities and intensity ratios, visualisation of data and log-transformation of ratios, global or local normalization of intensity ratios, and segmentation into different copy number regions using step detection algorithms.

- Class discovery analysis: This analytic approach, sometimes called unsupervised classification or knowledge discovery, tries to identify whether microarrays (objects, patients, mice, etc.) or genes cluster together in groups. Identifying naturally existing groups of objects (microarrays or genes) which cluster together can enable the discovery of new groups that otherwise were not previously known to exist. During knowledge discovery analysis, various unsupervised classification techniques can be employed with DNA microarray data to identify novel clusters (classes) of arrays. This type of approach is not hypothesis-driven, but rather is based on iterative pattern recognition or statistical learning methods to find an "optimal" number of clusters in the data. Examples of unsupervised analyses methods include self-organizing maps, neural gas, k-means cluster analyses, hierarchical cluster analysis, Genomic Signal Processing based clustering and model-based cluster analysis. For some of these methods the user also has to define a distance measure between pairs of objects. Although the Pearson correlation coefficient is usually employed, several other measures have been proposed and evaluated in the literature. The input data used in class discovery analyses are commonly based on lists of genes having high informativeness (low noise) based on low values of the coefficient of variation or high values of Shannon entropy, etc. The determination of the most likely or optimal number of clusters obtained from an unsupervised analysis is called cluster validity. Some commonly used metrics for cluster validity are the silhouette index, Davies-Bouldin index, Dunn's index, or Hubert's Γ statistic.

- Class prediction analysis: This approach, called supervised classification, establishes the basis for developing a predictive model into which future unknown test objects can be input in order to predict the most likely class membership of the test objects. Supervised analysis for class prediction involves use of techniques such as linear regression, k-nearest neighbor, learning vector quantization, decision tree analysis, random forests, naive Bayes, logistic regression, kernel regression, artificial neural networks, support vector machines, mixture of experts, and supervised neural gas. In addition, various metaheuristic methods are employed, such as genetic algorithms, covariance matrix self-adaptation, particle swarm optimization, and ant colony optimization. Input data for class prediction are usually based on filtered lists of genes which are predictive of class, determined using classical hypothesis tests, Gini diversity index, or information gain (entropy).

- Hypothesis-driven statistical analysis: Identification of statistically significant changes in gene expression are commonly identified using the t-test, ANOVA, Bayesian method Mann–Whitney test methods tailored to microarray data sets, which take into account multiple comparisons or cluster analysis. These methods assess statistical power based on the variation present in the data and the number of experimental replicates, and can help minimize Type I and type II errors in the analyses.

- Dimensional reduction: Analysts often reduce the number of dimensions (genes) prior to data analysis. This may involve linear approaches such as principal components analysis (PCA), or non-linear manifold learning (distance metric learning) using kernel PCA, diffusion maps, Laplacian eigenmaps, local linear embedding, locally preserving projections, and Sammon's mapping.

- Network-based methods: Statistical methods that take the underlying structure of gene networks into account, representing either associative or causative interactions or dependencies among gene products. Weighted gene co-expression network analysis is widely used for identifying co-expression modules and intramodular hub genes. Modules may corresponds to cell types or pathways. Highly connected intramodular hubs best represent their respective modules.

Microarray data may require further processing aimed at reducing the dimensionality of the data to aid comprehension and more focused analysis. Other methods permit analysis of data consisting of a low number of biological or technical replicates; for example, the Local Pooled Error (LPE) test pools standard deviations of genes with similar expression levels in an effort to compensate for insufficient replication.

Annotation

The relation between a probe and the mRNA that it is expected to detect is not trivial. Some mRNAs may cross-hybridize probes in the array that are supposed to detect another mRNA. In addition, mRNAs may experience amplification bias that is sequence or molecule-specific. Thirdly, probes that are designed to detect the mRNA of a particular gene may be relying on genomic EST information that is incorrectly associated with that gene.

Data Warehousing

Microarray data was found to be more useful when compared to other similar datasets. The sheer volume of data, specialized formats (such as MIAME), and curation efforts associated with the datasets require specialized databases to store the data. A number of open-source data warehousing solutions, such as InterMine and BioMart, have been created for the specific purpose of integrating diverse biological datasets, and also support analysis.

CELL CULTURE

Cell culture is the process by which cells are grown under controlled conditions, generally outside their natural environment. After the cells of interest have been isolated from living tissue, they can subsequently be maintained under carefully controlled conditions. These conditions vary for each cell type, but generally consist of a suitable vessel with a substrate or medium that supplies the essential nutrients (amino acids, carbohydrates, vitamins, minerals), growth factors, hormones, and gases (CO_2, O_2), and regulates the physio-chemical environment (pH buffer, osmotic pressure, temperature). Most cells require a surface or an artificial substrate (adherent or monolayer culture) whereas others can be grown free floating in culture medium (suspension culture). The lifespan of most cells is genetically determined, but some cell culturing cells have been "transformed" into immortal cells which will reproduce indefinitely if the optimal conditions are provided.

In practice, the term "cell culture" now refers to the culturing of cells derived from multicellular eukaryotes, especially animal cells, in contrast with other types of culture that also grow cells, such as plant tissue culture, fungal culture, and microbiological culture (of microbes). The historical development and methods of cell culture are closely interrelated to those of tissue culture and organ culture. Viral culture is also related, with cells as hosts for the viruses.

The laboratory technique of maintaining live cell lines (a population of cells descended from a single cell and containing the same genetic makeup) separated from their original tissue source became more robust in the middle 20th century.

Concepts in Mammalian Cell Culture

Isolation of Cells

Cells can be isolated from tissues for *ex vivo* culture in several ways. Cells can be easily purified from blood; however, only the white cells are capable of growth in culture. Cells can be isolated from solid tissues by digesting the extracellular matrix using enzymes such as collagenase, trypsin, or pronase, before agitating the tissue to release the cells into suspension. Alternatively, pieces of tissue can be placed in growth media, and the cells that grow out are available for culture. This method is known as explant culture.

Cells that are cultured directly from a subject are known as primary cells. With the exception of some derived from tumors, most primary cell cultures have limited lifespan.

An established or immortalized cell line has acquired the ability to proliferate indefinitely either through random mutation or deliberate modification, such as artificial expression of the telomerase gene. Numerous cell lines are well established as representative of particular cell types.

Maintaining Cells in Culture

For the majority of isolated primary cells, they undergo the process of senescence and stop dividing after a certain number of population doublings while generally retaining their viability (described as the Hayflick limit).

Cells are grown and maintained at an appropriate temperature and gas mixture (typically, 37 °C, 5% CO_2 for mammalian cells) in a cell incubator. Culture conditions vary widely for each cell type, and variation of conditions for a particular cell type can result in different phenotypes.

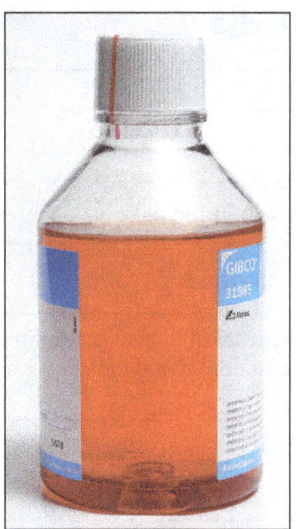

A bottle of DMEM cell culture medium.

Aside from temperature and gas mixture, the most commonly varied factor in culture systems is the cell growth medium. Recipes for growth media can vary in pH, glucose concentration, growth factors, and the presence of other nutrients. The growth factors used to supplement media are often derived from the serum of animal blood, such as fetal bovine serum (FBS), bovine calf serum, equine serum, and porcine serum. One complication of these blood-derived ingredients is the potential for contamination of the culture with viruses or prions, particularly in medical biotechnology applications. Current practice is to minimize or eliminate the use of these ingredients wherever possible and use human platelet lysate (hPL). This eliminates the worry of cross-species contamination when using FBS with human cells. hPL has emerged as a safe and reliable alternative as a direct replacement for FBS or other animal serum. In addition, chemically defined media can be used to eliminate any serum trace (human or animal), but this cannot always be accomplished with different cell types. Alternative strategies involve sourcing the animal blood from countries with minimum BSE/TSE risk, such as The United States, Australia and New Zealand, and using purified nutrient concentrates derived from serum in place of whole animal serum for cell culture.

Plating density (number of cells per volume of culture medium) plays a critical role for some cell types. For example, a lower plating density makes granulosa cells exhibit estrogen production, while a higher plating density makes them appear as progesterone-producing theca lutein cells.

Cells can be grown either in suspension or adherent cultures. Some cells naturally live in suspension, without being attached to a surface, such as cells that exist in the bloodstream. There are also cell lines that have been modified to be able to survive in suspension cultures so they can be grown to a higher density than adherent conditions would allow. Adherent cells require a surface, such as tissue culture plastic or microcarrier, which may be coated with extracellular matrix (such as collagen and laminin) components to increase adhesion properties and provide other signals needed for growth and differentiation. Most cells derived from solid tissues are adherent. Another type of adherent culture is organotypic culture, which involves growing cells in a three-dimensional (3-D) environment as opposed to two-dimensional culture dishes. This 3D culture system is biochemically and physiologically more similar to *in vivo* tissue, but is technically challenging to maintain because of many factors (e.g. diffusion).

Components of Cell Culture Media

Component	Function
Carbon source (glucose/glutamine)	Source of energy
Amino acid	Building blocks of protein
Vitamins	Promote cell survival and growth
Balanced salt solution	An isotonic mixture of ions to maintain optimum osmotic pressure within the cells and provide essential metal ions to act as cofactors for enzymatic reactions, cell adhesion etc.
Phenol red dye	pH indicator. The color of phenol red changes from orange/red at pH 7-7.4 to yellow at acidic (lower) pH and purple at basic (higher) pH
Bicarbonate /HEPES buffer	It is used to maintain a balanced pH in the media

Typical Growth Conditions

Parameter	
Temperature	37 °C
CO_2	5%
Relative Humidity	95%

Cell Line Cross-contamination

Cell line cross-contamination can be a problem for scientists working with cultured cells. Studies suggest anywhere from 15–20% of the time, cells used in experiments have been misidentified or contaminated with another cell line. Problems with cell line cross-contamination have even been detected in lines from the NCI-60 panel, which are used routinely for drug-screening studies. Major cell line repositories, including the American Type Culture Collection (ATCC), the European Collection of Cell Cultures (ECACC) and the German Collection of Microorganisms and Cell Cultures (DSMZ), have received cell line submissions from researchers that were misidentified by them. Such contamination poses a problem for the quality of research produced using cell culture lines, and the major repositories are now authenticating all cell line submissions. ATCC uses short tandem repeat (STR) DNA fingerprinting to authenticate its cell lines.

To address this problem of cell line cross-contamination, researchers are encouraged to authenticate their cell lines at an early passage to establish the identity of the cell line. Authentication should be repeated before freezing cell line stocks, every two months during active culturing and before any publication of research data generated using the cell lines. Many methods are used to identify cell lines, including isoenzyme analysis, human lymphocyte antigen (HLA) typing, chromosomal analysis, karyotyping, morphology and STR analysis.

One significant cell-line cross contaminant is the immortal HeLa cell line.

Other Technical Issues

As cells generally continue to divide in culture, they generally grow to fill the available area or volume. This can generate several issues:

- Nutrient depletion in the growth media.

- Changes in pH of the growth media.

- Accumulation of apoptotic/necrotic (dead) cells.

- Cell-to-cell contact can stimulate cell cycle arrest, causing cells to stop dividing, known as contact inhibition.

- Cell-to-cell contact can stimulate cellular differentiation.

- Genetic and epigenetic alterations, with a natural selection of the altered cells potentially leading to overgrowth of abnormal, culture-adapted cells with decreased differentiation and increased proliferative capacity.

Manipulation of Cultured Cells

Among the common manipulations carried out on culture cells are media changes, passaging cells, and transfecting cells. These are generally performed using tissue culture methods that rely on aseptic technique. Aseptic technique aims to avoid contamination with bacteria, yeast, or other cell lines. Manipulations are typically carried out in a biosafety cabinet or laminar flow cabinet to exclude contaminating micro-organisms. Antibiotics (e.g. penicillin and streptomycin) and anti-fungals (e.g. amphotericin B) can also be added to the growth media.

As cells undergo metabolic processes, acid is produced and the pH decreases. Often, a pH indicator is added to the medium to measure nutrient depletion.

Media Changes

In the case of adherent cultures, the media can be removed directly by aspiration, and then is replaced. Media changes in non-adherent cultures involve centrifuging the culture and resuspending the cells in fresh media.

Passaging Cells

Passaging (also known as subculture or splitting cells) involves transferring a small number of cells into a new vessel. Cells can be cultured for a longer time if they are split regularly, as it avoids the senescence associated with prolonged high cell density. Suspension cultures are easily passaged with a small amount of culture containing a few cells diluted in a larger volume of fresh media. For adherent cultures, cells first need to be detached; this is commonly done with a mixture of trypsin-EDTA; however, other enzyme mixes are now available for this purpose. A small number of detached cells can then be used to seed a new culture. Some cell cultures, such as RAW cells are mechanically scraped from the surface of their vessel with rubber scrapers.

Transfection and Transduction

Another common method for manipulating cells involves the introduction of foreign DNA by transfection. This is often performed to cause cells to express a gene of interest. More recently, the transfection of RNAi constructs have been realized as a convenient mechanism for suppressing the expression of a particular gene/protein. DNA can also be inserted into cells using viruses, in methods referred to as transduction, infection or transformation. Viruses, as parasitic agents, are well suited to introducing DNA into cells, as this is a part of their normal course of reproduction.

Established Human Cell Lines

Cell lines that originate with humans have been somewhat controversial in bioethics, as they may outlive their parent organism and later be used in the discovery of lucrative medical treatments.

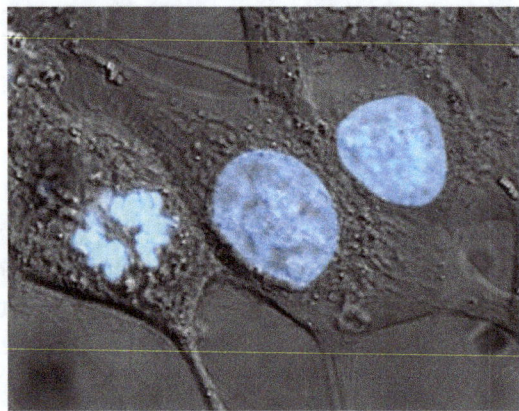

Cultured HeLa cells have been stained with Hoechst turning their nuclei blue, and are one of the earliest human cell lines descended from Henrietta Lacks, who died of cervical cancer from which these cells originated.

It is possible to fuse normal cells with an immortalised cell line. This method is used to produce monoclonal antibodies. In brief, lymphocytes isolated from the spleen (or possibly blood) of an immunised animal are combined with an immortal myeloma cell line (B cell lineage) to produce a hybridoma which has the antibody specificity of the primary lymphocyte and the immortality of the myeloma. Selective growth medium (HA or HAT) is used to select against unfused myeloma cells; primary lymphoctyes die quickly in culture and only the fused cells survive. These are screened for production of the required antibody, generally in pools to start with and then after single cloning.

Cell Strains

A cell strain is derived either from a primary culture or a cell line by the selection or cloning of cells having specific properties or characteristics which must be defined. Cell strains are cells that have been adapted to culture but, unlike cell lines, have a finite division potential. Non-immortalized cells stop dividing after 40 to 60 population doublings and, after this, they lose their ability to proliferate (a genetically determined event known as senescence).

Applications of Cell Culture

Mass culture of animal cell lines is fundamental to the manufacture of viral vaccines and other products of biotechnology. Culture of human stem cells is used to expand the number of cells and differentiate the cells into various somatic cell types for transplantation. Stem cell culture is also used to harvest the molecules and exosomes that the stem cells release for the purposes of therapeutic development.

Biological products produced by recombinant DNA (rDNA) technology in animal cell cultures include enzymes, synthetic hormones, immunobiologicals (monoclonal antibodies, interleukins, lymphokines), and anticancer agents. Although many simpler proteins can be produced using rDNA in bacterial cultures, more complex proteins that are glycosylated (carbohydrate-modified) currently must be made in animal cells. An important example of such a complex protein is the hormone erythropoietin. The cost of growing mammalian cell cultures is high, so research is underway to produce such complex proteins in insect cells or in higher plants, use of single embryonic cell and somatic embryos as a source for direct gene transfer via particle bombardment, transit gene expression and confocal microscopy observation is one of its applications. It also offers to confirm single cell origin of somatic embryos and the asymmetry of the first cell division, which starts the process.

Cell culture is also a key technique for cellular agriculture, which aims to provide both new products and new ways of producing existing agricultural products like milk, (cultured) meat, fragrances, and rhino horn from cells and microorganisms. It is therefore considered one means of achieving animal-free agriculture. It is also a central tool for teaching cell biology.

Cell Culture in Two Dimensions

Research in tissue engineering, stem cells and molecular biology primarily involves cultures of cells on flat plastic dishes. This technique is known as two-dimensional (2D) cell culture, and was first developed by Wilhelm Roux who, in 1885, removed a portion of the medullary plate of an embryonic chicken and maintained it in warm saline for several days on a flat glass plate. From the advance of polymer technology arose today's standard plastic dish for 2D cell culture, commonly known as the Petri dish. Julius Richard Petri, a German bacteriologist, is generally credited with this invention while working as an assistant to Robert Koch. Various researchers today also utilize culturing laboratory flasks, conicals, and even disposable bags like those used in single-use bioreactors.

Aside from Petri dishes, scientists have long been growing cells within biologically derived matrices such as collagen or fibrin, and more recently, on synthetic hydrogels such as polyacrylamide or PEG. They do this in order to elicit phenotypes that are not expressed on conventionally rigid substrates. There is growing interest in controlling matrix stiffness, a concept that has led to discoveries in fields such as:

- Stem cell self-renewal.
- Lineage specification.
- Cancer cell phenotype.
- Fibrosis.
- Hepatocyte function.
- Mechanosensing.

Cell Culture in Three Dimensions

Cell culture in three dimensions has been touted as "Biology's New Dimension". At present, the practice of cell culture remains based on varying combinations of single or multiple cell structures in 2D. Currently, there is an increase in use of 3D cell cultures in research areas including drug discovery, cancer biology, regenerative medicine and basic life science research. 3D cell cultures can be grown using a scaffold or matrix, or in a scaffold-free manner. Scaffold based cultures utilize an acellular 3D matrix or a liquid matrix. Scaffold-free methods are normally generated in suspensions. There are a variety of platforms used to facilitate the growth of three-dimensional cellular structures including scaffold systems such as hydrogel matrices and solid scaffolds, and scaffold-free systems such as low-adhesion plates, nanoparticle facilitated magnetic levitation, and hanging drop plates.

3D Cell Culture in Scaffolds

Eric Simon, in a 1988 NIH SBIR grant report, showed that electrospinning could be used to produced nano- and submicron-scale polystyrene and polycarbonate fibrous scaffolds specifically intended for use as *in vitro* cell substrates. This early use of electrospun fibrous lattices for cell culture and tissue engineering showed that various cell types including Human Foreskin Fibroblasts (HFF), transformed Human Carcinoma (HEp-2), and Mink Lung Epithelium (MLE) would adhere to and proliferate upon polycarbonate fibers. It was noted that, as opposed to the flattened morphology typically seen in 2D culture, cells grown on the electrospun fibers exhibited a more histotypic rounded 3-dimensional morphology generally observed in vivo.

3D Cell Culture in Hydrogels

As the natural extracellular matrix (ECM) is important in the survival, proliferation, differentiation and migration of cells, different hydrogel culture matrices mimicking natural ECM structure are seen as potential approaches to in vivo–like cell culturing. Hydrogels are composed of interconnected pores with high water retention, which enables efficient transport of substances such as nutrients and gases. Several different types of hydrogels from natural and synthetic materials are available for 3D cell culture, including animal ECM extract hydrogels, protein hydrogels, peptide hydrogels, polymer hydrogels, and wood-based nanocellulose hydrogel.

3D Cell Culturing by Magnetic Levitation

The 3D Cell Culturing by Magnetic Levitation method (MLM) is the application of growing 3D tissue by inducing cells treated with magnetic nanoparticle assemblies in spatially varying magnetic fields using neodymium magnetic drivers and promoting cell to cell interactions by levitating the cells up to the air/liquid interface of a standard petri dish. The magnetic nanoparticle assemblies consist of magnetic iron oxide nanoparticles, gold nanoparticles, and the polymer polylysine. 3D cell culturing is scalable, with the capability for culturing 500 cells to millions of cells or from single dish to high-throughput low volume systems.

Tissue Culture and Engineering

Cell culture is a fundamental component of tissue culture and tissue engineering, as it establishes the basics of growing and maintaining cells *in vitro*. The major application of human cell culture is in stem cell industry, where mesenchymal stem cells can be cultured and cryopreserved for future use. Tissue engineering potentially offers dramatic improvements in low cost medical care for hundreds of thousands of patients annually.

Vaccines

Vaccines for polio, measles, mumps, rubella, and chickenpox are currently made in cell cultures. Due to the H5N1 pandemic threat, research into using cell culture for influenza vaccines is being funded by the United States government. Novel ideas in the field include recombinant DNA-based vaccines, such as one made using human adenovirus (a common cold virus) as a vector, and novel adjuvants.

Culture of Non-mammalian Cells

Besides the culture of well-established immortalised cell lines, cells from primary explants of a plethora of organisms can be cultured for a limited period of time before sensecence occurs. Cultured primary cells have been extensively used in research, as is the case of fish keratocytes in cell migration studies.

Plant Cell Culture Methods

Plant cell cultures are typically grown as cell suspension cultures in a liquid medium or as callus cultures on a solid medium. The culturing of undifferentiated plant cells and calli requires the proper balance of the plant growth hormones auxin and cytokinin.

Insect Cell Culture

Cells derived from Drosophila melanogaster (most prominently, Schneider 2 cells) can be used for experiments which may be hard to do on live flies or larvae, such as biochemical studies or studies using siRNA. Cell lines derived from the army worm *Spodoptera frugiperda*, including Sf9 and Sf21, and from the cabbage looper *Trichoplusia ni*, High Five cells, are commonly used for expression of recombinant proteins using baculovirus.

Bacterial and Yeast Culture Methods

For bacteria and yeasts, small quantities of cells are usually grown on a solid support that contains nutrients embedded in it, usually a gel such as agar, while large-scale cultures are grown with the cells suspended in a nutrient broth.

Viral Culture Methods

The culture of viruses requires the culture of cells of mammalian, plant, fungal or bacterial origin as hosts for the growth and replication of the virus. Whole wild type viruses, recombinant viruses or viral products may be generated in cell types other than their natural hosts under the right conditions. Depending on the species of the virus, infection and viral replication may result in host cell lysis and formation of a viral plaque.

Common Cell Lines

Human cell lines:

- DU145 (prostate cancer).

- H295R (adrenocortical cancer).

- HeLa (cervical cancer).

- KBM-7 (chronic myelogenous leukemia).

- LNCaP (prostate cancer).

- MCF-7 (breast cancer).

- MDA-MB-468 (breast cancer).

- PC3 (prostate cancer).

- SaOS-2 (bone cancer).

- SH-SY5Y (neuroblastoma, cloned from a myeloma).

- T-47D (breast cancer).

- THP-1 (acute myeloid leukemia).

- U87 (glioblastoma).

- National Cancer Institute's 60 cancer cell line panel (NCI60).

Primate cell lines:

- Vero (African green monkey *Chlorocebus* kidney epithelial cell line).

Mouse cell lines:

- MC3T3 (embryonic calvarium).

Rat tumor cell lines:

- GH3 (pituitary tumor).

- PC12 (pheochromocytoma).

Plant cell lines:

- Tobacco BY-2 cells (kept as cell suspension culture, they are model system of plant cell).

Other species cell lines:

- Dog MDCK kidney epithelial.

- Xenopus A6 kidney epithelial.

- Zebrafish AB9.

TRANSFECTION

Transfection is the process of deliberately introducing naked or purified nucleic acids into eukaryotic cells. It may also refer to other methods and cell types, although other terms are often preferred: "transformation" is typically used to describe non-viral DNA transfer in bacteria and non-animal eukaryotic cells, including plant cells. In animal cells, transfection is the preferred term as transformation is also used to refer to progression to a cancerous state (carcinogenesis) in these cells. Transduction is often used to describe virus-mediated gene transfer into eukaryotic cells.

Genetic material (such as supercoiled plasmid DNA or siRNA constructs), or even proteins such as antibodies, may be transfected.

Transfection of animal cells typically involves opening transient pores or "holes" in the cell membrane to allow the uptake of material. Transfection can be carried out using calcium phosphate (i.e. tricalcium phosphate), by electroporation, by cell squeezing or by mixing a cationic lipid with the material to produce liposomes that fuse with the cell membrane and deposit their cargo inside.

Transfection can result in unexpected morphologies and abnormalities in target cells.

Methods

There are various methods of introducing foreign DNA into a eukaryotic cell: some rely on physical treatment (electroporation, cell squeezing, nanoparticles, magnetofection); others rely on chemical materials or biological particles (viruses) that are used as carriers. Gene delivery is, for example, one of the steps necessary for gene therapy and the genetic modification of crops. There are many different methods of gene delivery developed for various types of cells and tissues, from bacterial to mammalian. Generally, the methods can be divided into two categories: non-viral and viral.

Non-viral methods include physical methods such as electroporation, microinjection, gene gun, impalefection, hydrostatic pressure, continuous infusion, and sonication and chemical, such as lipofection, which is a lipid-mediated DNA-transfection process utilizing liposome vectors. It can also include the use of polymeric gene carriers (polyplexes).

Virus mediated gene delivery utilizes the ability of a virus to inject its DNA inside a host cell. A gene that is intended for delivery is packaged into a replication-deficient viral particle. Viruses used to date include retrovirus, lentivirus, adenovirus, adeno-associated virus, and herpes simplex virus. However, there are drawbacks to using viruses to deliver genes into cells. Viruses can only deliver very small pieces of DNA into the cells, it is labor-intensive and there are risks of random insertion sites, cytopathic effects and mutagenesis.

Nonviral Methods

Chemical-based Transfection

Chemical-based transfection can be divided into several kinds: cyclodextrin, polymers, liposomes, or nanoparticles:

- One of the cheapest methods uses calcium phosphate, originally discovered by F. L. Graham and A. J. van der Eb in 1973. HEPES-buffered saline solution (HeBS) containing phosphate ions is combined with a calcium chloride solution containing the DNA to be transfected. When the two are combined, a fine precipitate of the positively charged calcium and the negatively charged phosphate will form, binding the DNA to be transfected on its surface. The suspension of the precipitate is then added to the cells to be transfected (usually a cell culture grown in a monolayer). By a process not entirely understood, the cells take up some of the precipitate, and with it, the DNA. This process has been a preferred method of identifying many oncogenes.

- Other methods use highly branched organic compounds, so-called dendrimers, to bind the DNA and get it into the cell.

- Another method is the use of cationic polymers such as DEAE-dextran or polyethylenimine (PEI). The negatively charged DNA binds to the polycation and the complex is taken up by the cell via endocytosis.

- Lipofection (or liposome transfection) is a technique used to inject genetic material into a cell by means of liposomes, which are vesicles that can easily merge with the cell membrane since they are both made of a phospholipid bilayer. Lipofection generally uses a positively charged (cationic) lipid (cationic liposomes or mixtures) to form an aggregate with the negatively charged (anionic) genetic material. This transfection technology performs the same tasks as other biochemical procedures utilizing polymers, DEAE-dextran, calcium phosphate, and electroporation. The efficiency of lipofection can be improved by treating transfected cells with a mild heat shock.

- Fugene is a series of widely used proprietary non-liposomal transfection reagents capable of directly transfecting a wide variety of cells with high efficiency and low toxicity.

Non-chemical Methods

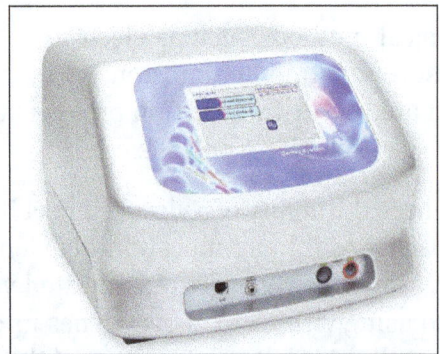

Electroporator with square wave and exponential decay waveforms for in vitro, in vivo, adherent cell and 96 well electroporation applications.

- Electroporation (gene electrotransfer) is a popular method, where transient increase in the permeability of cell membrane is achieved when the cells are exposed to short pulses of an intense electric field.

- Cell squeezing is a method invented in 2012 by Armon Sharei, Robert Langer and Klavs Jensen at MIT. It enables delivery of molecules into cells via cell membrane deformation. It is a high throughput vector-free microfluidic platform for intracellular delivery. It reduces the possibility of toxicity or off-target effects as it does not rely on exogenous materials or electrical fields.

- Sonoporation uses high-intensity ultrasound to induce pore formation in cell membranes. This pore formation is attributed mainly to the cavitation of gas bubbles interacting with nearby cell membranes since it is enhanced by the addition of ultrasound contrast agent, a source of cavitation nuclei.

- Optical transfection is a method where a tiny (~1 µm diameter) hole is transiently generated in the plasma membrane of a cell using a highly focused laser. This technique was first described in 1984 by Tsukakoshi et al., who used a frequency tripled Nd:YAG to generate stable and transient transfection of normal rat kidney cells. In this technique, one cell at a time is treated, making it particularly useful for single cell analysis.

- Protoplast fusion is a technique in which transformed bacterial cells are treated with lysozyme in order to remove the cell wall. Following this, fusogenic agents (e.g., Sendai virus, PEG, electroporation) are used in order to fuse the protoplast carrying the gene of interest with the target recipient cell. A major disadvantage of this method is that bacterial components are non-specifically introduced into the target cell as well.

- Impalefection is a method of introducing DNA bound to a surface of a nanofiber that is inserted into a cell. This approach can also be implemented with arrays of nanofibers that are introduced into large numbers of cells and intact tissue.

- Hydrodynamic delivery is a method used in mice and rats, but to a lesser extent in larger animals, in which DNA most often in plasmids (including transposons) can be delivered to the liver using hydrodynamic injection that involves infusion of a relatively large volume in the blood in less than 10 seconds; nearly all of the DNA is expressed in the liver by this procedure.

Particle-based Methods

- A direct approach to transfection is the gene gun, where the DNA is coupled to a nanoparticle of an inert solid (commonly gold), which is then "shot" directly into the target cell's nucleus.

- Magnetofection, or magnet-assisted transfection, is a transfection method that uses magnetic force to deliver DNA into target cells. Nucleic acids are first associated with magnetic nanoparticles. Then, application of magnetic force drives the nucleic acid particle complexes towards and into the target cells, where the cargo is released.

- Impalefection is carried out by impaling cells by elongated nanostructures and arrays of such nanostructures such as carbon nanofibers or silicon nanowires that have been functionalized with plasmid DNA.

- Another particle-based method of transfection is known as particle bombardment. The nucleic acid is delivered through membrane penetration at a high velocity, usually connected to microprojectiles.

Other Hybrid Methods

Other methods of transfection include nucleofection, which has proved very efficient in transfection of the THP-1 cell line, creating a viable cell line that was able to be differentiated into mature macrophages, and heat shock.

Viral Methods

DNA can also be introduced into cells using viruses as a carrier. In such cases, the technique is called viral transduction, and the cells are said to be transduced. Adenoviral vectors can be useful for viral transfection methods because they can transfer genes into a wide variety of human cells and have high transfer rates. Lentiviral and vectors are also helpful due to their ability to transduce cells not currently undergoing mitosis.

Stable and Transient Transfection

Stable and transient transfection differ in their long term effects on a cell; a stably-transfected cell will continuously express transfected DNA and pass it on to daughter cells, while a transiently-transfected cell will express transfected DNA for a short amount of time and not pass it on to daughter cells.

For some applications of transfection, it is sufficient if the transfected genetic material is only transiently expressed. Since the DNA introduced in the transfection process is usually not integrated into the nuclear genome, the foreign DNA will be diluted through mitosis or degraded. Cell lines expressing the Epstein–Barr virus (EBV) nuclear antigen 1 (EBNA1) or the SV40 large-T antigen, allow episomal amplification of plasmids containing the viral EBV (293E) or SV40 (293T) origins of replication, greatly reducing the rate of dilution.

If it is desired that the transfected gene actually remain in the genome of the cell and its daughter cells, a stable transfection must occur. To accomplish this, a marker gene is co-transfected, which gives the cell some selectable advantage, such as resistance towards a certain toxin. Some (very few) of the transfected cells will, by chance, have integrated the foreign genetic material into their genome. If the toxin is then added to the cell culture, only those few cells with the marker gene integrated into their genomes will be able to proliferate, while other cells will die. After applying this selective stress (selection pressure) for some time, only the cells with a stable transfection remain and can be cultivated further.

Common agents for selecting stable transfection are:

- Geneticin, or G418, neutralized by the product of the neomycin resistance gene.

- Puromycin.

- Zeocin.

- Hygromycin B.

- Blasticidin S.

RNA Transfection

RNA can also be transfected into cells to transiently express its coded protein, or to study RNA decay kinetics. RNA transfection is often used in primary cells that do not divide.

siRNAs can also be transfected to achieve RNA silencing (i.e. loss of RNA and protein from the targeted gene). This has become a major application in research to achieve "knock-down" of proteins of interests (e.g. Endothelin-1) with potential applications in gene therapy. Limitation of the silencing approach are the toxicity of the transfection for cells and potential "off-target" effects on the expression of other genes/proteins.

References

- Gel-electrophoresis, dna-sequencing-pcr-electrophoresis, biotech-dna-technology, biology, science: khanacademy.org: Retrieved 23 January, 2019

- Hartl, Daniel L., Jones, Elizabeth W. (2001), Genetics: Analysis of Genes and Genomes, Fifth Edition. ISBN 0-7637-0913-1

- Polymerase-chain-reaction-pcr, dna-sequencing-pcr-electrophoresis, biotech-dna-technology, biology, science: khanacademy.org, Retrieved 24 February, 2019

- Behjati S, Tarpey PS (December 2013). "What is next generation sequencing?". Archives of Disease in Childhood. Education and Practice Edition. 98 (6): 236–8. Doi:10.1136/archdischild-2013-304340. PMC 3841808. PMID 23986538

- Chiosea, SI; Williams, L; Griffith, CC; Thompson, LD; Weinreb, I; Bauman, JE; Luvison, A; Roy, S; Seethala, RR; Nikiforova, MN (June 2015). "Molecular characterization of apocrine salivary duct carcinoma". The American Journal of Surgical Pathology. 39 (6): 744–52

- Molecular-Cloning, life-sciences: news-medical.net, Retrieved 25 March, 2019

- Venetia A. Saunders (6 December 2012). Microbial genetics applied to biotechnology :: principles and techniques of Gene Transfer and Manipulation. Springer. ISBN 9781461597964

- Overview-dna-cloning, dna-cloning-tutorial, biotech-dna-technology, biology, science: khanacademy.org, Retrieved 26 April, 2019

- Kwok, Albert; Eggimann, Gabriela A.; Heitz, Marc; Reymond, Jean-Louis; Hollfelder, Florian; Darbre, Tamis (2016-11-09). "Efficient Transfection of sirna by Peptide Dendrimer-Lipid Conjugates". Chembiochem. 17 (23): 2223–2229. Doi:10.1002/cbic.201600485. ISSN 1439-4227

Diverse Aspects in Molecular Biology

5 CHAPTER

Some of the fundamental concepts studied in molecular biology include DNA and RNA replication, protein biosynthesis, oligonucleotide synthesis, genome and proteome. This chapter closely examines these fundamental concepts of molecular biology to provide an extensive understanding of the subject.

DNA REPLICATION

DNA is the genetic material that defines every cell. Before a cell duplicates and is divided into new daughter cells through either mitosis or meiosis, biomolecules and organelles must be copied to be distributed among the cells. DNA, found within the nucleus, must be replicated in order to ensure that each new cell receives the correct number of chromosomes. The process of DNA duplication is called DNA replication. Replication follows several steps that involve multiple proteins called replication enzymes and RNA. In eukaryotic cells, such as animal cells and plant cells, DNA replication occurs in the S phase of interphase during the cell cycle. The process of DNA replication is vital for cell growth, repair, and reproduction in organisms.

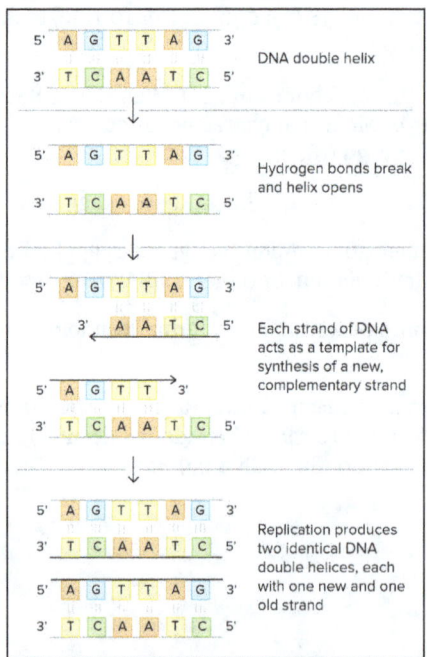

There are about 3 billion base pairs of DNA in your genome, all of which must be accurately copied when any one of your trillions of cells divides. The basic mechanisms of DNA replication are similar across organisms.

DNA replication is semiconservative, meaning that each strand in the DNA double helix acts as a template for the synthesis of a new, complementary strand.

This process takes us from one starting molecule to two "daughter" molecules, with each newly formed double helix containing one new and one old strand. In a sense, that's all there is to DNA replication. But what's actually most interesting about this process is how it's carried out in a cell.

Cells need to copy their DNA very quickly, and with very few errors (or risk problem such as cancer). To do so, they use a variety of enzymes and proteins, which work together to make sure DNA replication is performed smoothly and accurately.

DNA Polymerase

One of the key molecules in DNA replication is the enzyme DNA polymerase. DNA polymerases are responsible for synthesizing DNA: they add nucleotides one by one to the growing DNA chain, incorporating only those that are complementary to the template.

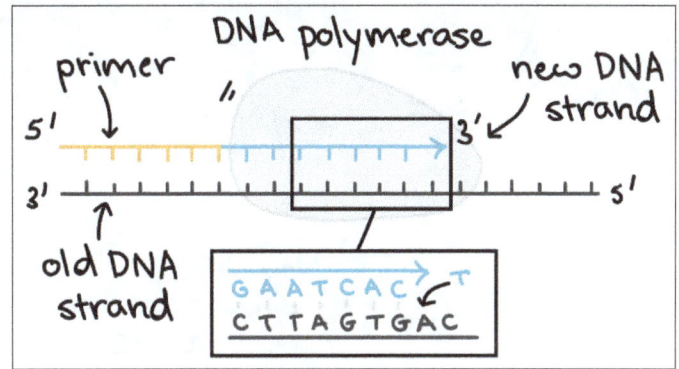

Here are some key features of DNA polymerases:

- They always need a template.

- They can only add nucleotides to the 3' end of a DNA strand.

- They can't start making a DNA chain from scratch, but require a pre-existing chain or short stretch of nucleotides called a primer.

- They proofread, or check their work, removing the vast majority of "wrong" nucleotides that are accidentally added to the chain.

The addition of nucleotides requires energy. This energy comes from the nucleotides themselves, which have three phosphates attached to them (much like the energy-carrying molecule ATP). When the bond between phosphates is broken, the energy released is used to form a bond between the incoming nucleotide and the growing chain.

In prokaryotes such as E. coli, there are two main DNA polymerases involved in DNA replication: DNA pol III (the major DNA-maker), and DNA pol I, which plays a crucial supporting role.

Starting DNA Replication

How do DNA polymerases and other replication factors know where to begin? Replication always starts at specific locations on the DNA, which are called origins of replication and are recognized by their sequence.

E. coli, like most bacteria, has a single origin of replication on its chromosome. The origin is about 245 base pairs long and has mostly A/T base pairs (which are held together by fewer hydrogen bonds than G/C base pairs), making the DNA strands easier to separate.

Specialized proteins recognize the origin, bind to this site, and open up the DNA. As the DNA opens, two Y-shaped structures called replication forks are formed, together making up what's called a replication bubble. The replication forks will move in opposite directions as replication proceeds.

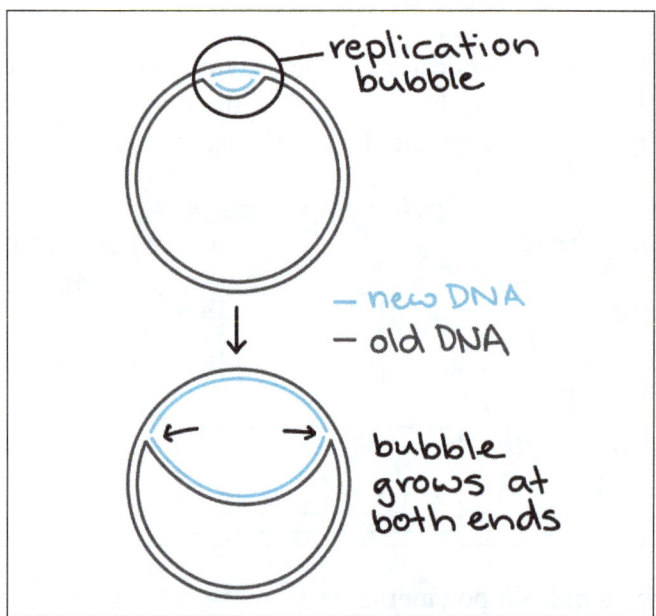

Diagram based on similar illustration in Reece et al.

How does replication actually get going at the forks? Helicase is the first replication enzyme to load on at the origin of replication. Helicase's job is to move the replication forks forward by "unwinding" the DNA (breaking the hydrogen bonds between the nitrogenous base pairs).

Proteins called single-strand binding proteins coat the separated strands of DNA near the replication fork, keeping them from coming back together into a double helix.

Primers and Primase

DNA polymerases can only add nucleotides to the 3' end of an existing DNA strand. (They use the free -OH group found at the 3' end as a "hook," adding a nucleotide to this group in the polymerization reaction.) How, then, does DNA polymerase add the first nucleotide at a new replication fork?

The problem is solved with the help of an enzyme called primase. Primase makes an RNA primer, or short stretch of nucleic acid complementary to the template, that provides a 3' end for DNA polymerase to work on. A typical primer is about five to ten nucleotides long. The primer primes DNA synthesis, i.e., gets it started.

Once the RNA primer is in place, DNA polymerase "extends" it, adding nucleotides one by one to make a new DNA strand that's complementary to the template strand.

Leading and Lagging Strands

In E. coli, the DNA polymerase that handles most of the synthesis is DNA polymerase III. There are two molecules of DNA polymerase III at a replication fork, each of them hard at work on one of the two new DNA strands.

DNA polymerases can only make DNA in the 5' to 3' direction, and this poses a problem during replication. A DNA double helix is always anti-parallel; in other words, one strand runs in the 5' to 3' direction, while the other runs in the 3' to 5' direction. This makes it necessary for the two new strands, which are also antiparallel to their templates, to be made in slightly different ways.

One new strand, which runs 5' to 3' towards the replication fork, is the easy one. This strand is made continuously, because the DNA polymerase is moving in the same direction as the replication fork. This continuously synthesized strand is called the leading strand.

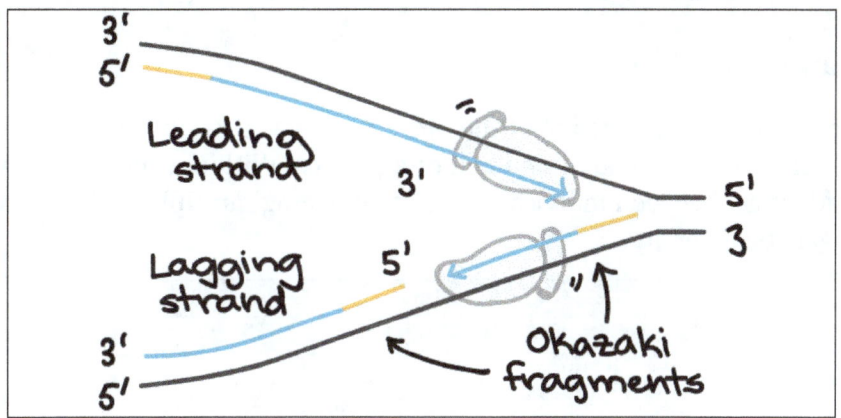

The other new strand, which runs 5' to 3' away from the fork, is trickier. This strand is made in fragments because, as the fork moves forward, the DNA polymerase (which is moving away from the fork) must come off and reattach on the newly exposed DNA. This tricky strand, which is made in fragments, is called the lagging strand.

The small fragments are called Okazaki fragments, named for the Japanese scientist who discovered them. The leading strand can be extended from one primer alone, whereas the lagging strand needs a new primer for each of the short Okazaki fragments.

Maintenance and Cleanup Crew

Some other proteins and enzymes, in addition the main ones above, are needed to keep DNA replication running smoothly. One is a protein called the sliding clamp, which holds DNA polymerase III molecules in place as they synthesize DNA. The sliding clamp is a ring-shaped protein and keeps the DNA polymerase of the lagging strand from floating off when it re-starts at a new Okazaki fragment.

Topoisomerase also plays an important maintenance role during DNA replication. This enzyme prevents the DNA double helix ahead of the replication fork from getting too tightly wound as the DNA is opened up. It acts by making temporary nicks in the helix to release the tension, then sealing the nicks to avoid permanent damage.

Finally, there is a little cleanup work to do if we want DNA that doesn't contain any RNA or gaps. The RNA primers are removed and replaced by DNA through the activity of DNA polymerase I, the other polymerase involved in replication. The nicks that remain after the primers are replaced get sealed by the enzyme DNA ligase.

GENOME

In the fields of molecular biology and genetics, a genome is the genetic material of an organism. It consists of DNA (or RNA in RNA viruses). The genome includes both the genes (the coding regions) and the noncoding DNA, as well as mitochondrial DNA and chloroplast DNA. The study of the genome is called genomics.

Sequencing and Mapping

A genome sequence is the complete list of the nucleotides (A, C, G, and T for DNA genomes) that make up all the chromosomes of an individual or a species. Within a species, the vast majority of nucleotides are identical between individuals, but sequencing multiple individuals is necessary to understand the genetic diversity.

Part of DNA sequence - prototypification of complete genome of virus.

In 1976, Walter Fiers at the University of Ghent (Belgium) was the first to establish the complete nucleotide sequence of a viral RNA-genome (Bacteriophage MS2). The next year, Fred Sanger completed the first DNA-genome sequence: Phage Φ-X174, of 5386 base pairs. The first complete genome sequences among all three domains of life were released within a short period during the mid-1990s: The first bacterial genome to be sequenced was that of Haemophilus influenzae, completed by a team at The Institute for Genomic Research in 1995. A few months later, the first eukaryotic genome was completed, with sequences of the 16 chromosomes of budding yeast *Saccharomyces cerevisiae* published as the result of a European-led effort begun in the mid-1980s. The first genome sequence for an archaeon, *Methanococcus jannaschii*, was completed in 1996, again by The Institute for Genomic Research.

The development of new technologies has made genome sequencing dramatically cheaper and easier, and the number of complete genome sequences is growing rapidly. The US National Institutes of Health maintains one of several comprehensive databases of genomic information. Among the thousands of completed genome sequencing projects include those for rice, a mouse, the plant *Arabidopsis thaliana*, the puffer fish, and the bacteria E. coli. In December 2013, scientists first sequenced the entire *genome* of a Neanderthal, an extinct species of humans. The genome was extracted from the toe bone of a 130,000-year-old Neanderthal found in a Siberian cave.

New sequencing technologies, such as massive parallel sequencing have also opened up the prospect of personal genome sequencing as a diagnostic tool, as pioneered by Manteia Predictive Medicine. A major step toward that goal was the completion in 2007 of the full genome of James D. Watson, one of the co-discoverers of the structure of DNA.

Whereas a genome sequence lists the order of every DNA base in a genome, a genome map identifies the landmarks. A genome map is less detailed than a genome sequence and aids in navigating around the genome. The Human Genome Project was organized to map and to sequence the human genome. A fundamental step in the project was the release of a detailed genomic map by Jean Weissenbach and his team at the Genoscope in Paris.

Reference genome sequences and maps continue to be updated, removing errors and clarifying regions of high allelic complexity. The decreasing cost of genomic mapping has permitted genealogical sites to offer it as a service, to the extent that one may submit one's genome to crowdsourced scientific endeavours.

Viral Genomes

Viral genomes can be composed of either RNA or DNA. The genomes of RNA viruses can be either single-stranded or double-stranded RNA, and may contain one or more separate RNA molecules. DNA viruses can have either single-stranded or double-stranded genomes. Most DNA virus genomes are composed of a single, linear molecule of DNA, but some are made up of a circular DNA molecule.

Prokaryotic Genomes

Prokaryotes and eukaryotes have DNA genomes. Archaea have a single circular chromosome. Most bacteria also have a single circular chromosome; however, some bacterial species have linear chromosomes or multiple chromosomes. If the DNA is replicated faster than the bacterial cells divide, multiple copies of the chromosome can be present in a single cell, and if the cells divide faster than the DNA can be replicated, multiple replication of the chromosome is initiated before the division occurs, allowing daughter cells to inherit complete genomes and already partially replicated chromosomes. Most prokaryotes have very little repetitive DNA in their genomes. However, some symbiotic bacteria (e.g. *Serratia symbiotica*) have reduced genomes and a high fraction of pseudogenes: only ~40% of their DNA encodes proteins.

Some bacteria have auxiliary genetic material, also part of their genome, which is carried in plasmids. For this, the word *genome* should not be used as a synonym of chromosome.

Eukaryotic Genomes

Eukaryotic genomes are composed of one or more linear DNA chromosomes. The number of chromosomes varies widely from Jack jumper ants and an asexual nemotode, which each have only one pair, to a fern species that has 720 pairs. A typical human cell has two copies of each of 22 autosomes, one inherited from each parent, plus two sex chromosomes, making it diploid. Gametes, such as ova, sperm, spores, and pollen, are haploid, meaning they carry only one copy of each chromosome.

In addition to the chromosomes in the nucleus, organelles such as the chloroplasts and mitochondria have their own DNA. Mitochondria are sometimes said to have their own genome often referred to as the "mitochondrial genome". The DNA found within the chloroplast may be referred to as the "plastome". Like the bacteria they originated from, mitochondria and chloroplasts have a circular chromosome.

Unlike prokaryotes, eukaryotes have exon-intron organization of protein coding genes and variable amounts of repetitive DNA. In mammals and plants, the majority of the genome is composed of repetitive DNA.

Coding Sequences

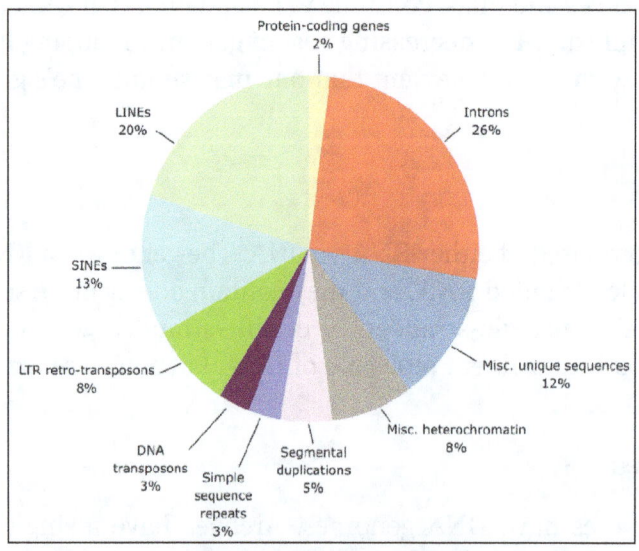

Composition of the human genome.

DNA sequences that carry the instructions to make proteins are coding sequences. The proportion of the genome occupied by coding sequences varies widely. A larger genome does not necessarily contain more genes, and the proportion of non-repetitive DNA decreases along with increasing genome size in complex eukaryotes.

Simple eukaryotes such as *C. elegans* and fruit fly, have more non-repetitive DNA than repetitive DNA, while the genomes of more complex eukaryotes tend to be composed largely of repetitive DNA. In some plants and amphibians, the proportion of repetitive DNA is more than 80%. Similarly, only 2% of the human genome codes for proteins.

Noncoding Sequences

Noncoding sequences include introns, sequences for non-coding RNAs, regulatory regions, and repetitive DNA. Noncoding sequences make up 98% of the human genome. There are two categories of repetitive DNA in the genome: tandem repeats and interspersed repeats.

Tandem Repeats

Short, non-coding sequences that are repeated head-to-tail are called tandem repeats. Microsatellites consisting of 2-5 basepair repeats, while minisatellite repeats are 30-35 bp. Tandem repeats make up about 4% of the human genome and 9% of the fruit fly genome. Tandom repeats can be functional. For example, telomeres are composed of the tandem repeat TTAGGG in mammals, and they play an important role in protecting the ends of the chromosome.

In other cases, expansions in the number of tandem repeats in exons or introns can cause disease. For example, the human gene huntingtin typically contains 6–29 tandem repeats of the nucleotides CAG (encoding a polyglutamine tract). An expansion to over 36 repeats results in Huntington's disease, a neurodegenerative disease. Twenty human disorders are known to result from similar tandem repeat expansions in various genes. The mechanism by which proteins with expanded polygulatamine tracts cause death of neurons is not fully understood. One possibility is that the proteins fail to fold properly and avoid degradation, instead accumulating in aggregates that also sequester important transcription factors, thereby altering gene expression.

Tandem repeats are usually caused by slippage during replication, unequal crossing-over and gene conversion.

Transposable Elements

Transposable elements (TEs) are sequences of DNA with a defined structure that are able to change their location in the genome. TEs are categorized as either class I TEs, which replicate by a copy-and-paste mechanism, or class II TEs, which can be excised from the genome and inserted at a new location.

The movement of TEs is a driving force of genome evolution in eukaryotes because their insertion can disrupt gene functions, homologous recombination between TEs can produce duplications, and TE can shuffle exons and regulatory sequences to new locations.

Retrotransposons

Retrotransposons can be transcribed into RNA, which are then duplicated at another site into the genome. Retrotransposons can be divided into Long terminal repeats (LTRs) and Non-Long Terminal Repeats (Non-LTR).

Long terminal repeats (LTRs) are derived from ancient retroviral infections, so they encode proteins related to retroviral proteins including gag (structural proteins of the virus), pol (reverse transcriptase and integrase), pro (protease), and in some cases env (envelope) genes. These genes are flanked by long repeats at both 5' and 3' ends. It has been reported that LTRs consist of the largest fraction in most plant genome and might account for the huge variation in genome size.

Non-long terminal repeats (Non-LTRs) are classified as long interspersed elements (LINEs), short interspersed elements (SINEs), and Penelope-like elements. In *Dictyostelium discoideum*, there is another DIRS-like elements belong to Non-LTRs. Non-LTRs are widely spread in eukaryotic genomes.

Long interspersed elements (LINEs) encode genes for reverse transcriptase and endonuclease, making them autonomous transposable elements. The human genome has around 500,000 LINEs, taking around 17% of the genome.

Short interspersed elements (SINEs) are usually less than 500 base pairs and are non-autonomous, so they rely on the proteins encoded by LINEs for transposition. The Alu element is the most common SINE found in primates. It is about 350 base pairs and occupies about 11% of the human genome with around 1,500,000 copies.

DNA Transposons

DNA transposons encode a transposase enzyme between inverted terminal repeats. When expressed, the transposase recognizes the terminal inverted repeats that flank the transposon and catalyzes its excision and reinsertion in a new site. This cut-and-paste mechanism typically reinserts transposons near their original location (within 100kb). DNA transposons are found in bacteria and make up 3% of the human genome and 12% of the genome of the roundworm *C. elegans*.

Genome size

Genome size is the total number of DNA base pairs in one copy of a haploid genome. In humans, the nuclear genome comprises approximately 3.2 billion nucleotides of DNA, divided into 24 linear molecules, the shortest 50 000 000 nucleotides in length and the longest 260 000 000 nucleotides, each contained in a different chromosome. The genome size is positively correlated with the morphological complexity among prokaryotes and lower eukaryotes; however, after mollusks and all the other higher eukaryotes above, this correlation is no longer effective. This phenomenon also indicates the mighty influence coming from repetitive DNA on the genomes.

Since genomes are very complex, one research strategy is to reduce the number of genes in a genome to the bare minimum and still have the organism in question survive. There is experimental work being done on minimal genomes for single cell organisms as well as minimal genomes for multi-cellular organisms. The work is both in vivo and in silico.

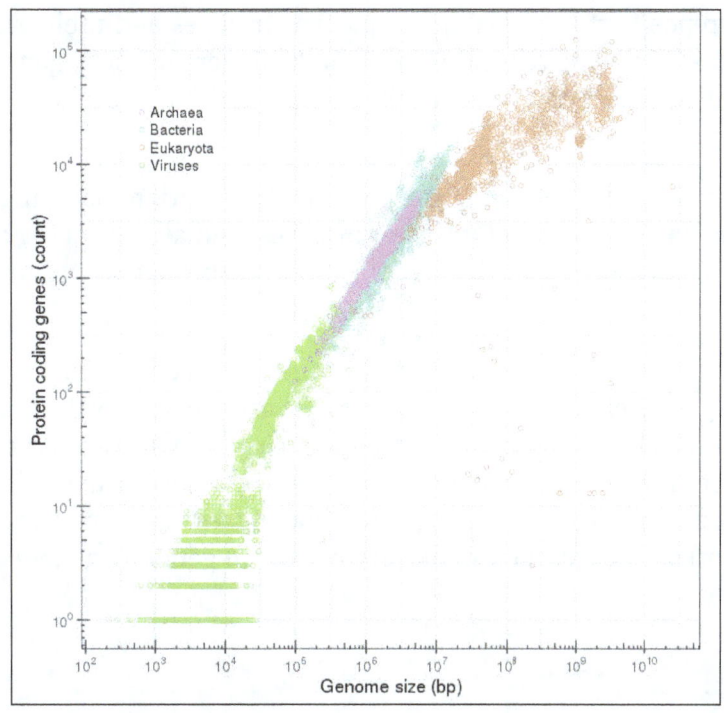

Log-log plot of the total number of annotated proteins in genomes submitted to GenBank as a function of genome size.

Genomic Alterations

All the cells of an organism originate from a single cell, so they are expected to have identical genomes; however, in some cases, differences arise. Both the process of copying DNA during cell division and exposure to environmental mutagens can result in mutations in somatic cells. In some cases, such mutations lead to cancer because they cause cells to divide more quickly and invade surrounding tissues. In certain lymphocytes in the human immune system, V(D)J recombination generates different genomic sequences such that each cell produces a unique antibody or T cell receptors.

During meiosis, diploid cells divide twice to produce haploid germ cells. During this process, recombination results in a reshuffling of the genetic material from homologous chromosomes so each gamete has a unique genome.

Genome-wide Reprogramming

Genome-wide reprogramming in mouse primordial germ cells involves epigenetic imprint erasure leading to totipotency. Reprogramming is facilitated by active DNA demethylation, a process that entails the DNA base excision repair pathway. This pathway is employed in the erasure of CpG methylation (5mC) in primordial germ cells. The erasure of 5mC occurs via its conversion to 5-hydroxymethylcytosine (5hmC) driven by high levels of the ten-eleven dioxygenase enzymes TET1 and TET2.

Genome Evolution

Genomes are more than the sum of an organism's genes and have traits that may be measured and studied without reference to the details of any particular genes and their products. Researchers compare traits such as karyotype (chromosome number), genome size, gene order, codon usage bias, and GC-content to determine what mechanisms could have produced the great variety of genomes that exist today.

Duplications play a major role in shaping the genome. Duplication may range from extension of short tandem repeats, to duplication of a cluster of genes, and all the way to duplication of entire chromosomes or even entire genomes. Such duplications are probably fundamental to the creation of genetic novelty.

Horizontal gene transfer is invoked to explain how there is often an extreme similarity between small portions of the genomes of two organisms that are otherwise very distantly related. Horizontal gene transfer seems to be common among many microbes. Also, eukaryotic cells seem to have experienced a transfer of some genetic material from their chloroplast and mitochondrial genomes to their nuclear chromosomes. Recent empirical data suggest an important role of viruses and sub-viral RNA-networks to represent a main driving role to generate genetic novelty and natural genome editing.

PROTEOME

The proteome is the entire set of proteins that is, or can be, expressed by a genome, cell, tissue, or organism at a certain time. It is the set of expressed proteins in a given type of cell or organism, at a given time, under defined conditions. Proteomics is the study of the proteome.

Systems

The term has been applied to several different types of biological systems. A cellular proteome is the collection of proteins found in a particular cell type under a particular set of environmental conditions such as exposure to hormone stimulation. It can also be useful to consider an organism's complete proteome, which can be conceptualized as the complete set of proteins from all of the various cellular proteomes. This is very roughly the protein equivalent of the genome. The term "proteome" has also been used to refer to the collection of proteins in certain sub-cellular biological systems. For example, all of the proteins in a virus can be called a viral proteome.

Size and Contents

The proteome can be larger than the genome, especially in eukaryotes, as more than one protein can be produced from one gene due to alternative splicing (e.g. human proteome consists 92,179 proteins out of which 71,173 are splicing variants). On the other hand, not all genes are translated to proteins, and many known genes encode only RNA which is the final functional product. Moreover, complete proteome size vary depending the kingdom of life. For instance, eukaryotes, bacteria, archaea and viruses have on average 15,145, 3,200, 2,358 and 42 proteins respectively encoded in their genomes.

The term "dark proteome" coined by Perdigão and colleagues, defines regions of proteins that have no detectable sequence homology to other proteins of known three-dimensional structure and therefore cannot be modeled by homology. For 546,000 Swiss-Prot proteins, 44–54% of the proteome in eukaryotes and viruses was found to be "dark", compared with only ~14% in archaea and bacteria.

Methods to Study the Proteome

Numerous methods are available to study proteins, sets of proteins, or the whole proteome. In fact, proteins are often studied indirectly, e.g. using computational methods and analyses of genomes.

Separation Techniques and Electrophoresis

Proteomics, the study of the proteome, has largely been practiced through the separation of proteins by two dimensional gel electrophoresis. In the first dimension, the proteins are separated by isoelectric focusing, which resolves proteins on the basis of charge. In the second dimension, proteins are separated by molecular weight using SDS-PAGE. The gel is dyed with Coomassie Brilliant Blue or silver to visualize the proteins. Spots on the gel are proteins that have migrated to specific locations.

Mass Spectrometry

Mass spectrometry has augmented proteomics. Peptide mass fingerprinting identifies a protein by cleaving it into short peptides and then deduces the protein's identity by matching the observed peptide masses against a sequence database. Tandem mass spectrometry, on the other hand, can get sequence information from individual peptides by isolating them, colliding them with a non-reactive gas, and then cataloguing the fragment ions produced.

In May 2014, a draft map of the human proteome was published. This map was generated using high-resolution Fourier-transform mass spectrometry. This study profiled 30 histologically normal human samples resulting in the identification of proteins coded by 17,294 genes. This accounts for around 84% of the total annotated protein-coding genes.

Protein Complementation Assays and Interaction Screens

Protein fragment complementation assays are often used to detect protein–protein interactions. The yeast two-hybrid assay is the most popular of them but there are numerous variations, both used in vitro and in vivo.

PROTEIN BIOSYNTHESIS

Protein synthesis is the process whereby biological cells generate new proteins; it is balanced by the loss of cellular proteins via degradation or export. Translation, the assembly of amino acids by ribosomes, is an essential part of the biosynthetic pathway, along with generation of messenger RNA (mRNA), aminoacylation of transfer RNA (tRNA), cotranslational transport, and post-translational modification. Protein biosynthesis is strictly regulated at multiple steps. They are principally during transcription (phenomena of RNA synthesis from DNA template) and translation (phenomena of amino acid assembly from RNA).

The cistron DNA is transcribed into the first of a series of RNA intermediates. The last version is used as a template in synthesis of a polypeptide chain. Protein will often be synthesized directly from genes by translating mRNA. However, when a protein must be available on short notice or in large quantities, a protein precursor is produced. A proprotein is an inactive protein containing one or more inhibitory peptides that can be activated when the inhibitory sequence is removed by proteolysis during posttranslational modification. A preprotein is a form that contains a signal sequence (an N-terminal signal peptide) that specifies its insertion into or through membranes, i.e., targets them for secretion. The signal peptide is cleaved off in the endoplasmic reticulum. Preproproteins have both sequences (inhibitory and signal) still present.

In protein synthesis, a succession of tRNA molecules charged with appropriate amino acids are brought together with an mRNA molecule and matched up by base-pairing through the anti-codons of the tRNA with successive codons of the mRNA. The amino acids are then linked together to extend the growing protein chain, and the tRNAs, no longer carrying amino acids, are released. This whole complex of processes is carried out by the ribosome, formed of two main chains of RNA, called ribosomal RNA (rRNA), and more than 50 different proteins. The ribosome latches onto the end of an mRNA molecule and moves along it, capturing loaded tRNA molecules and joining together their amino acids to form a new protein chain.

Protein biosynthesis, although very similar, is different for prokaryotes and eukaryotes.

Transcription

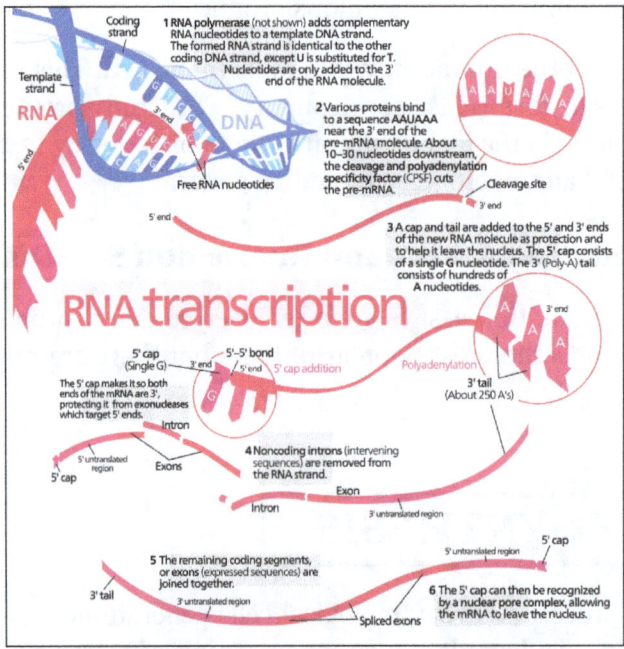

Diagram showing the process of transcription

In transcription an mRNA chain is generated, with one strand of the DNA double helix in the genome as a template. This strand is called the template strand. Transcription can be divided into 3 stages: initiation, elongation, and termination, each regulated by a large number of proteins such as transcription factors and coactivators that ensure that the correct gene is transcribed.

Transcription occurs in the cell nucleus, where the DNA is held and is never able to leave. The DNA structure of the cell is made up of two helixes made up of sugar and phosphate held together by hydrogen bonds between the bases of opposite strands. The sugar and the phosphate in each strand are joined together by stronger phosphodiester covalent bonds. The DNA is "unzipped" (disruption of hydrogen bonds between different single strands) by the enzyme helicase, leaving the single nucleotide chain open to be copied. RNA polymerase reads the DNA strand from the 3-prime (3') end to the 5-prime (5') end, while it synthesizes a single strand of messenger RNA in the 5'-to-3' direction. The general RNA structure is very similar to the DNA structure, but in RNA the nucleotide uracil takes the place that thymine occupies in DNA. The single strand of mRNA leaves the nucleus through nuclear pores, and migrates into the cytoplasm.

The first product of transcription differs in prokaryotic cells from that of eukaryotic cells, as in prokaryotic cells the product is mRNA, which needs no post-transcriptional modification, whereas, in eukaryotic cells, the first product is called primary transcript, that needs post-transcriptional modification (capping with 7-methyl-guanosine, tailing with a poly A tail) to give hnRNA (heterogeneous nuclear RNA). hnRNA then undergoes splicing of introns (noncoding parts of the gene) via spliceosomes to produce the final mRNA.

Translation

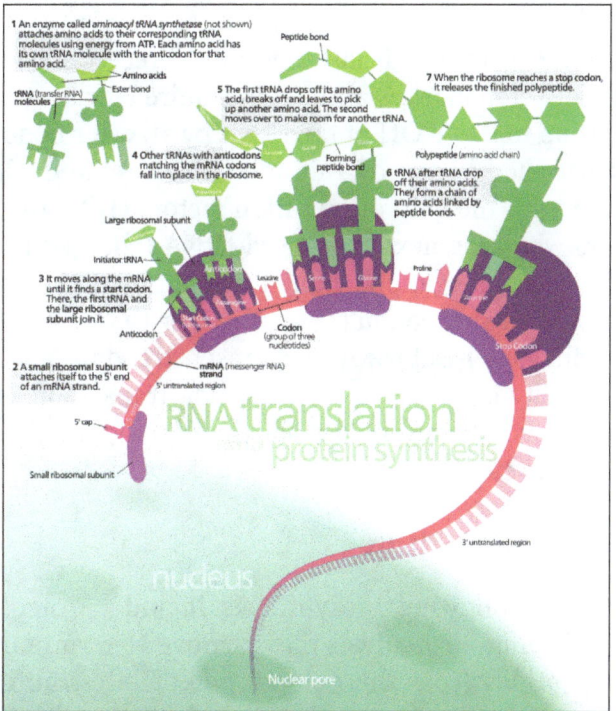

Diagram showing the process of translation.

Phenomena of amino acid assembly from RNA. The synthesis of proteins from RNA is known as translation. In eukaryotes, translation occurs in the cytoplasm, where the ribosomes are located. Ribosomes are made of a small and large subunit that surround the mRNA. In translation, messenger RNA (mRNA) is decoded to produce a specific polypeptide according to the rules specified by the trinucleotide genetic code. This uses an mRNA sequence as a template to guide the synthesis of a chain of amino acids that form a protein. Translation proceeds in four phases: activation, initiation, elongation, and termination (all describing the growth of the amino acid chain, or polypeptide that is the product of translation).

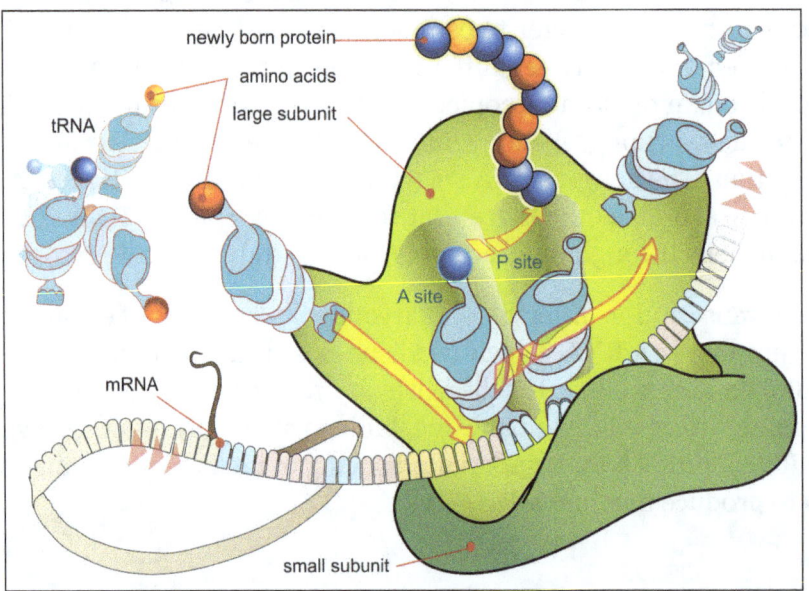

Diagram showing the translation of mRNA and the synthesis of proteins by a ribosome.

In activation, the correct amino acid (AA) is joined to the correct transfer RNA (tRNA). While this is not, in the technical sense, a step in translation, it is required for translation to proceed. The AA is joined by its carboxyl group to the 3' OH of the tRNA by an ester bond. When the tRNA has an amino acid linked to it, it is termed "charged". Initiation involves the small subunit of the ribosome binding to 5' end of mRNA with the help of initiation factors (IF), other proteins that assist the process. Elongation occurs when the next aminoacyl-tRNA (charged tRNA) in line binds to the ribosome along with GTP and an elongation factor. Termination of the polypeptide happens when the A site of the ribosome faces a stop codon (UAA, UAG, or UGA). When this happens, no tRNA can recognize it, but releasing factor can recognize nonsense codons and causes the release of the polypeptide chain. The capacity of disabling or inhibiting translation in protein biosynthesis is used by some antibiotics such as anisomycin, cycloheximide, chloramphenicol, tetracycline, streptomycin, erythromycin, puromycin, etc.

Events During Protein Translation

Events that occur during or following biosynthesis include proteolysis, post-translational modification and protein folding. Proteolysis may remove N-terminal, C-terminal or internal amino-acid residues or peptides from the polypeptide. The termini and side-chains of the polypeptide may be subjected to post-translational modification. These modifications may be required for correct cellular localisation or the natural function of the protein. During and after synthesis, polypeptide chains often fold to assume, so called, native secondary and tertiary structures. This is known as *protein folding* and is typically required for the natural function of the protein.

OLIGONUCLEOTIDE SYNTHESIS

Oligonucleotide synthesis is the chemical synthesis of relatively short fragments of nucleic acids with defined chemical structure (sequence). The technique is extremely useful in current laboratory practice because it provides a rapid and inexpensive access to custom-made oligonucleotides of the desired sequence. Whereas enzymes synthesize DNA and RNA only in a 5' to 3' direction, chemical oligonucleotide synthesis does not have this limitation, although it is, most often, carried out in the opposite, 3' to 5' direction. Currently, the process is implemented as solid-phase synthesis using phosphoramidite method and phosphoramidite building blocks derived from protected 2'-deoxynucleosides (dA, dC, dG, and T), ribonucleosides (A, C, G, and U), or chemically modified nucleosides, e.g. LNA or BNA.

To obtain the desired oligonucleotide, the building blocks are sequentially coupled to the growing oligonucleotide chain in the order required by the sequence of the product. The process has been fully automated since the late 1970s. Upon the completion of the chain assembly, the product is released from the solid phase to solution, deprotected, and collected. The occurrence of side reactions sets practical limits for the length of synthetic oligonucleotides (up to about 200 nucleotide residues) because the number of errors accumulates with the length of the oligonucleotide being synthesized. Products are often isolated by high-performance liquid chromatography (HPLC) to obtain the desired oligonucleotides in high purity. Typically, synthetic oligonucleotides are single-stranded DNA or RNA molecules around 15–25 bases in length.

Oligonucleotides find a variety of applications in molecular biology and medicine. They are most commonly used as antisense oligonucleotides, small interfering RNA, primers for DNA sequencing and amplification, probes for detecting complementary DNA or RNA via molecular hybridization, tools for the targeted introduction of mutations and restriction sites, and for the synthesis of artificial genes.

Synthesis by the Phosphoramidite Method

Nucleoside Phosphoramidites

Protected 2'-deoxynucleoside phosphoramidites.

The naturally occurring nucleotides (nucleoside-3'- or 5'-phosphates) and their phosphodiester analogs are insufficiently reactive to afford an expeditious synthetic preparation of oligonucleotides in high yields. The selectivity and the rate of the formation of internucleosidic linkages is dramatically improved by using 3'-*O*-(*N,N*-diisopropyl phosphoramidite) derivatives of nucleosides (nucleoside phosphoramidites) that serve as building blocks in phosphite triester methodology. To prevent undesired side reactions, all other functional groups present in nucleosides have to be rendered unreactive (protected) by attaching protecting groups. Upon the completion of the oligonucleotide chain assembly, all the protecting groups are removed to yield the desired oligonucleotides. Below, the protecting groups currently used in commercially available and most common nucleoside phosphoramidite building blocks are briefly reviewed:

- The 5'-hydroxyl group is protected by an acid-labile DMT (4,4'-dimethoxytrityl) group.

- Thymine and uracil, nucleic bases of thymidine and uridine, respectively, do not have exocyclic amino groups and hence do not require any protection.

- Although the nucleic base of guanosine and 2'-deoxyguanosine does have an exocyclic amino group, its basicity is low to an extent that it does not react with phosphoramidites under the conditions of the coupling reaction. However, a phosphoramidite derived from the N2-unprotected 5'-O-DMT-2'-deoxyguanosine is poorly soluble in acetonitrile, the solvent commonly used in oligonucleotide synthesis. In contrast, the N2-protected versions of the same compound dissolve in acetonitrile well and hence are widely used. Nucleic bases adenine and cytosine bear the exocyclic amino groups reactive with the activated phosphoramidites under the conditions of the coupling reaction. By the use of additional steps in the synthetic cycle or alternative coupling agents and solvent systems, the oligonucleotide chain assembly may be carried out using dA and dC phosphoramidites with unprotected amino groups. However, these approaches currently remain in the research stage. In routine oligonucleotide synthesis, exocyclic amino groups in nucleosides are kept permanently protected over the entire length of the oligonucleotide chain assembly.

The protection of the exocyclic amino groups has to be orthogonal to that of the 5'-hydroxy group because the latter is removed at the end of each synthetic cycle. The simplest to implement, and hence the most widely used, strategy is to install a base-labile protection group on the exocyclic amino groups. Most often, two protection schemes are used.

- In the first, the standard and more robust scheme, Bz (benzoyl) protection is used for A, dA, C, and dC, while G and dG are protected with isobutyryl group. More recently, Ac (acetyl) group is used to protect C and dC as shown in Figure.

- In the second, mild protection scheme, A and dA are protected with isobutyryl or phenoxyacetyl groups (PAC). C and dC bear acetyl protection, and G and dG are protected with 4-isopropylphenoxyacetyl (iPr-PAC) or dimethylformamidino (dmf) groups. Mild protecting groups are removed more readily than the standard protecting groups. However, the phosphoramidites bearing these groups are less stable when stored in solution.

- The phosphite group is protected by a base-labile 2-cyanoethyl group. Once a phosphoramidite has been coupled to the solid support-bound oligonucleotide and the phosphite moieties have been converted to the P(V) species, the presence of the phosphate protection is not mandatory for the successful conducting of further coupling reactions.

2'-*O*-protected ribonucleoside phosphoramidites.

- In RNA synthesis, the 2'-hydroxy group is protected with TBDMS (t-butyldimethylsilyl) group. or with TOM (tri-iso-propylsilyloxymethyl) group, both being removable by treatment with fluoride ion.

- The phosphite moiety also bears a diisopropylamino (iPr2N) group reactive under acidic conditions. Upon activation, the diisopropylamino group leaves to be substituted by the 5'-hydroxy group of the support-bound oligonucleotide.

Non-nucleoside Phosphoramidites

Non-nucleoside phosphoramidites for 5'-modification of synthetic oligonucleotides.
MMT = mono-methoxytrityl,(4-methoxyphenyl)diphenylmethyl.

Non-nucleoside phosphoramidites are the phosphoramidite reagents designed to introduce various functionalities at the termini of synthetic oligonucleotides or between nucleotide residues in the middle of the sequence. In order to be introduced inside the sequence, a non-nucleosidic modifier has to possess at least two hydroxy groups, one of which is often protected with the DMT group while the other bears the reactive phosphoramidite moiety.

Non-nucleosidic phosphoramidites are used to introduce desired groups that are not available in natural nucleosides or that can be introduced more readily using simpler chemical designs. A very short selection of commercial phosphoramidite reagents is shown in figure for the demonstration of the available structural and functional diversity. These reagents serve for the attachment of 5'-terminal phosphate (1), NH_2 (2), SH (3), aldehydo (4), and carboxylic groups (5), CC triple bonds (6), non-radioactive labels and quenchers (exemplified by 6-FAM amidite (7) for the attachment of fluorescein and dabcyl amidite (8), respectively), hydrophilic and hydrophobic modifiers (exemplified by hexaethyleneglycol amidite (9) and cholesterol amidite (10), respectively), and biotin amidite (11).

Synthetic Cycle

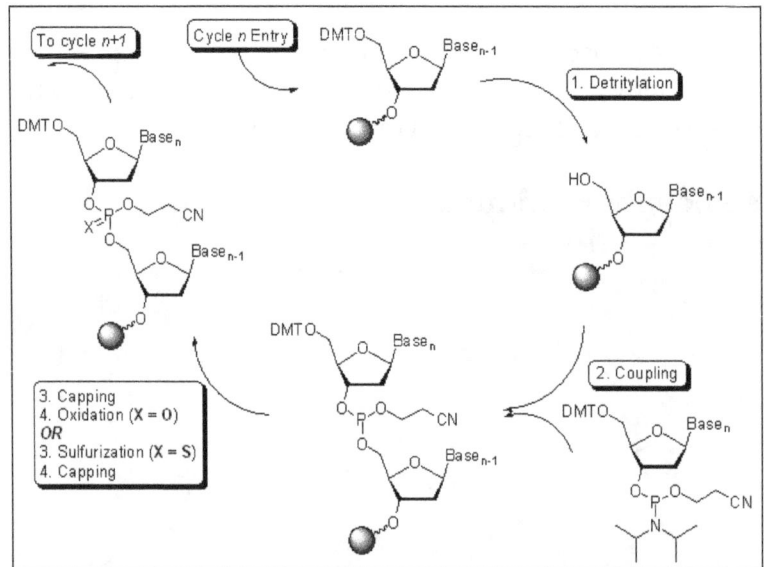

Synthetic cycle for preparation of oligonucleotides by phosphoramidite method.

Oligonucleotide synthesis is carried out by a stepwise addition of nucleotide residues to the 5'-terminus of the growing chain until the desired sequence is assembled. Each addition is referred to as a synthetic cycle and consists of four chemical reactions:

Step 1: De-blocking (detritylation)

The DMT group is removed with a solution of an acid, such as 2% trichloroacetic acid (TCA) or 3% dichloroacetic acid (DCA), in an inert solvent (dichloromethane or toluene). The orange-colored DMT cation formed is washed out; the step results in the solid support-bound oligonucleotide precursor bearing a free 5'-terminal hydroxyl group. It is worth remembering that conducting detritylation for an extended time or with stronger than recommended solutions of acids leads to depurination of solid support-bound oligonucleotide and thus reduces the yield of the desired full-length product.

Step 2: Coupling

A 0.02–0.2 M solution of nucleoside phosphoramidite (or a mixture of several phosphoramidites) in acetonitrile is activated by a 0.2–0.7 M solution of an acidic azole catalyst, $1H$-tetrazole,

5-ethylthio-1H-tetrazole, 2-benzylthiotetrazole, 4,5-dicyanoimidazole, or a number of similar compounds. A more extensive information on the use of various coupling agents in oligonucleotide synthesis can be found in a recent review. The mixing is usually very brief and occurs in fluid lines of oligonucleotide synthesizers while the components are being delivered to the reactors containing solid support. The activated phosphoramidite in 1.5 – 20-fold excess over the support-bound material is then brought in contact with the starting solid support (first coupling) or a support-bound oligonucleotide precursor (following couplings) whose 5'-hydroxy group reacts with the activated phosphoramidite moiety of the incoming nucleoside phosphoramidite to form a phosphite triester linkage. The coupling of 2'-deoxynucleoside phosphoramidites is very rapid and requires, on small scale, about 20 s for its completion. In contrast, sterically hindered 2'-O-protected ribonucleoside phosphoramidites require 5-15 min to be coupled in high yields. The reaction is also highly sensitive to the presence of water, particularly when dilute solutions of phosphoramidites are used, and is commonly carried out in anhydrous acetonitrile. Generally, the larger the scale of the synthesis, the lower the excess and the higher the concentration of the phosphoramidites is used. In contrast, the concentration of the activator is primarily determined by its solubility in acetonitrile and is irrespective of the scale of the synthesis. Upon the completion of the coupling, any unbound reagents and by-products are removed by washing.

Step 3: Capping

The capping step is performed by treating the solid support-bound material with a mixture of acetic anhydride and 1-methylimidazole or, less often, DMAP as catalysts and, in the phosphoramidite method, serves two purposes.

- After the completion of the coupling reaction, a small percentage of the solid support-bound 5'-OH groups (0.1 to 1%) remains unreacted and needs to be permanently blocked from further chain elongation to prevent the formation of oligonucleotides with an internal base deletion commonly referred to as (n-1) shortmers. The unreacted 5'-hydroxy groups are, to a large extent, acetylated by the capping mixture.

- It has also been reported that phosphoramidites activated with 1H-tetrazole react, to a small extent, with the O6 position of guanosine. Upon oxidation with I2 /water, this side product, possibly via O6-N7 migration, undergoes depurination. The apurinic sites thus formed are readily cleaved in the course of the final deprotection of the oligonucleotide under the basic conditions to give two shorter oligonucleotides thus reducing the yield of the full-length product. The O6 modifications are rapidly removed by treatment with the capping reagent as long as the capping step is performed prior to oxidation with I2/water.

- The synthesis of oligonucleotide phosphorothioates does not involve the oxidation with I2/water, and, respectively, does not suffer from the side reaction described above. On the other hand, if the capping step is performed prior to sulfurization, the solid support may contain the residual acetic anhydride and N-methylimidazole left after the capping step. The capping mixture interferes with the sulfur transfer reaction, which results in the extensive formation of the phosphate triester internucleosidic linkages in place of the desired PS triesters. Therefore, for the synthesis of OPS, it is advisable to conduct the sulfurization step prior to the capping step.

Step 4: Oxidation

The newly formed tricoordinated phosphite triester linkage is not natural and is of limited stability under the conditions of oligonucleotide synthesis. The treatment of the support-bound material with iodine and water in the presence of a weak base (pyridine, lutidine, or collidine) oxidizes the phosphite triester into a tetracoordinated phosphate triester, a protected precursor of the naturally occurring phosphate diester internucleosidic linkage. Oxidation may be carried out under anhydrous conditions using tert-Butyl hydroperoxide or, more efficiently, (1S)-(+)-(10-camphorsulfonyl)-oxaziridine (CSO). The step of oxidation may be substituted with a sulfurization step to obtain oligonucleotide phosphorothioates. In the latter case, the sulfurization step is best carried out prior to capping.

Solid Supports

In solid-phase synthesis, an oligonucleotide being assembled is covalently bound, via its 3'-terminal hydroxy group, to a solid support material and remains attached to it over the entire course of the chain assembly. The solid support is contained in columns whose dimensions depend on the scale of synthesis and may vary between 0.05 mL and several liters. The overwhelming majority of oligonucleotides are synthesized on small scale ranging from 10 nmol to 1 µmol. More recently, high-throughput oligonucleotide synthesis where the solid support is contained in the wells of multi-well plates (most often, 96 or 384 wells per plate) became a method of choice for parallel synthesis of oligonucleotides on small scale. At the end of the chain assembly, the oligonucleotide is released from the solid support and is eluted from the column or the well.

Solid Support Material

In contrast to organic solid-phase synthesis and peptide synthesis, the synthesis of oligonucleotides proceeds best on non-swellable or low-swellable solid supports. The two most often used solid-phase materials are controlled pore glass (CPG) and macroporous polystyrene (MPPS).

- CPG is commonly defined by its pore size. In oligonucleotide chemistry, pore sizes of 500, 1000, 1500, 2000, and 3000 Å are used to allow the preparation of about 50, 80, 100, 150, and 200-mer oligonucleotides, respectively. To make native CPG suitable for further processing, the surface of the material is treated with (3-aminopropyl)triethoxysilane to give aminopropyl CPG. The aminopropyl arm may be further extended to result in long chain aminoalkyl (LCAA) CPG. The amino group is then used as an anchoring point for linkers suitable for oligonucleotide synthesis.

- MPPS suitable for oligonucleotide synthesis is a low-swellable, highly cross-linked polystyrene obtained by polymerization of divinylbenzene (min 60%), styrene, and 4-chloromethylstyrene in the presence of a porogeneous agent. The macroporous chloromethyl MPPS obtained is converted to aminomethyl MPPS.

Linker Chemistry

To make the solid support material suitable for oligonucleotide synthesis, non-nucleosidic linkers or nucleoside succinates are covalently attached to the reactive amino groups in aminopropyl CPG, LCAA CPG, or aminomethyl MPPS. The remaining unreacted amino groups are capped with acetic anhydride. Typically, three conceptually different groups of solid supports are used.

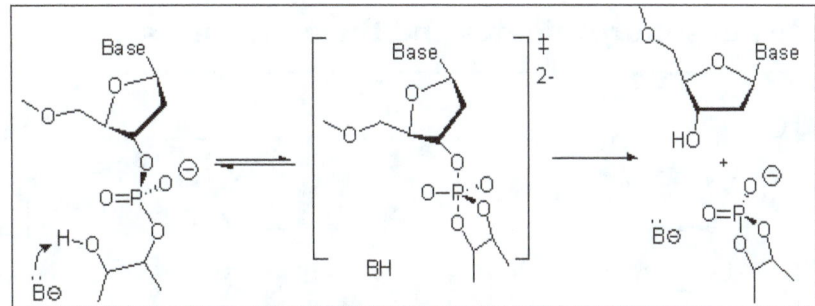

Commercial solid supports for oligonucleotide synthesis.

Mechanism of 3'-dephosphorylation of oligonucleotides
assembled on universal solid supports.

- Universal supports. In a more recent, more convenient, and more widely used method, the synthesis starts with the universal support where a non-nucleosidic linker is attached to the solid support material (compounds 1 and 2). A phosphoramidite respective to the 3'-terminal nucleoside residue is coupled to the universal solid support in the first synthetic cycle of oligonucleotide chain assembly using the standard protocols. The chain assembly is then continued until the completion, after which the solid support-bound oligonucleotide is deprotected. The characteristic feature of the universal solid supports is that the release of the oligonucleotides occurs by the hydrolytic cleavage of a P-O bond that attaches the 3'-O of the 3'-terminal nucleotide residue to the universal linker as shown in figure. The critical advantage of this approach is that the same solid support is used irrespectively of the sequence of the oligonucleotide to be synthesized. For the complete removal of the linker and the 3'-terminal phosphate from the assembled oligonucleotide, the solid support 1 and several similar solid supports require gaseous ammonia, aqueous ammonium hydroxide, aqueous methylamine, or their mixture and are commercially available. The solid support 2 requires a solution of ammonia in anhydrous methanol and is also commercially available.

- Nucleosidic solid supports. In a historically first and still popular approach, the 3'-hydroxy group of the 3'-terminal nucleoside residue is attached to the solid support via, most often, 3'-O-succinyl arm as in compound 3. The oligonucleotide chain assembly starts with the coupling of a phosphoramidite building block respective to the nucleotide residue second from the 3'-terminus. The 3'-terminal hydroxy group in oligonucleotides synthesized on nucleosidic solid supports is deprotected under the conditions somewhat milder than those applicable for universal solid supports. However, the fact that a nucleosidic solid support has to be selected in a sequence-specific manner reduces the throughput of the entire synthetic process and increases the likelihood of human error.

- Special solid supports are used for the attachment of desired functional or reporter groups at the 3'-terminus of synthetic oligonucleotides. For example, the commercial solid support 4 allows the preparation of oligonucleotides bearing 3'-terminal 3-aminopropyl linker. Similarly to non-nucleosidic phosphoramidites, many other special solid supports designed for the attachment of reactive functional groups, non-radioactive reporter groups, and terminal modifiers (e.c. cholesterol or other hydrophobic tethers) and suited for various applications are commercially available. A more detailed information on various solid supports for oligonucleotide synthesis can be found in a recent review.

Oligonucleotide Phosphorothioates and their Synthesis

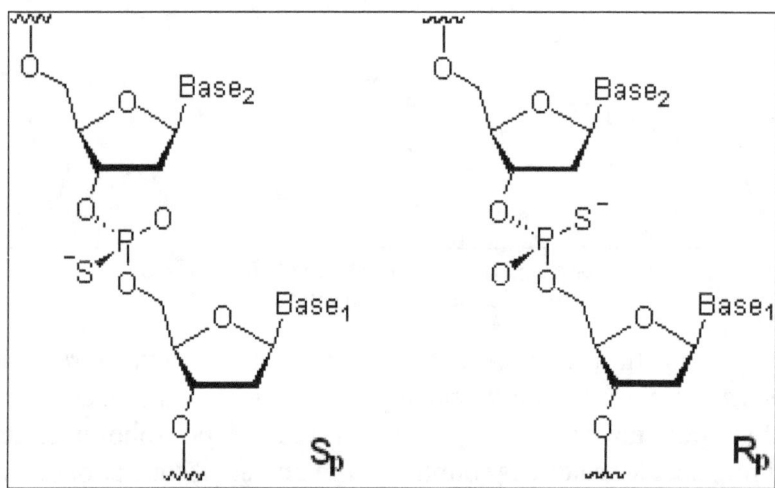

S_p and R_p-diastereomeric internucleosidic phosphorothioate linkages.

Oligonucleotide phosphorothioates (OPS) are modified oligonucleotides where one of the oxygen atoms in the phosphate moiety is replaced by sulfur. Only the phosphorothioates having sulfur at a non-bridging position as shown in figure are widely used and are available commercially. The replacement of the non-bridging oxygen with sulfur creates a new center of chirality at phosphorus. In a simple case of a dinucleotide, this results in the formation of a diastereomeric pair of S_p- and R_p-dinucleoside monophosphorothioates whose structures are shown in Figure. In an n-mer oligonucleotide where all $(n-1)$ internucleosidic linkages are phosphorothioate linkages, the number of diastereomers m is calculated as $m = 2^{(n-1)}$. Being non-natural analogs of nucleic acids, OPS are substantially more stable towards hydrolysis by nucleases, the class of enzymes that destroy nucleic acids by breaking the bridging P-O bond of the phosphodiester moiety. This property determines the use of OPS as antisense oligonucleotides in *in vitro* and *in vivo* applications where the extensive exposure to nucleases is inevitable. Similarly, to improve the stability of siRNA, at least one phosphorothioate linkage is often introduced at the 3'-terminus of both sense and antisense strands. In chirally pure OPS, all-Sp diastereomers are more stable to enzymatic degradation than their all-Rp analogs. However, the preparation of chirally pure OPS remains a synthetic challenge. In laboratory practice, mixtures of diastereomers of OPS are commonly used.

Synthesis of OPS is very similar to that of natural oligonucleotides. The difference is that the oxidation step is replaced by sulfur transfer reaction (sulfurization) and that the capping step is performed after the sulfurization. Of many reported reagents capable of the efficient sulfur transfer, only three are commercially available:

Commercial sulfur transfer agents for oligonucleotide synthesis.

- 3-(Dimethylaminomethylidene)amino-3H-1,2,4-dithiazole-3-thione, DDTT (3) provides rapid kinetics of sulfurization and high stability in solution. The reagent is available from several sources.

- 3H-1,2-benzodithiol-3-one 1,1-dioxide (4) also known as Beaucage reagent displays a better solubility in acetonitrile and short reaction times. However, the reagent is of limited stability in solution and is less efficient in sulfurizing RNA linkages.

- N,N,N'N'-Tetraethylthiuram disulfide (TETD) is soluble in acetonitrile and is commercially available. However, the sulfurization reaction of an internucleosidic DNA linkage with TETD requires 15 min, which is more than 10 times as slow as that with compounds 3 and 4.

Automation

In the past, oligonucleotide synthesis was carried out manually in solution or on solid phase. The solid phase synthesis was implemented using, as containers for the solid phase, miniature glass columns similar in their shape to low-pressure chromatography columns or syringes equipped with porous filters. Currently, solid-phase oligonucleotide synthesis is carried out automatically using computer-controlled instruments (oligonucleotide synthesizers) and is technically implemented in column, multi-well plate, and array formats. The column format is best suited for research and large scale applications where a high-throughput is not required. Multi-well plate format is designed specifically for high-throughput synthesis on small scale to satisfy the growing demand of industry and academia for synthetic oligonucleotides. A number of oligonucleotide synthesizers for small scale synthesis and medium to large scale synthesis are available commercially.

First Commercially Available Oligonucleotide Synthesizers

In March 1982 a practical course was hosted by the Department of Biochemistry, Technische Hochschule Darmstadt, Germany. M.H. Caruthers, M.J. Gait, H.G. Gassen, H.Koster, K. Itakura, and C. Birr among others attended. The program comprised practical work, lectures, and seminars on solid-phase chemical synthesis of oligonucleotides. A select group of 15 students attended and had an unprecedented opportunity to be instructed by the esteemed teaching staff.

Oligonucleotide Synthesizer.

Along with manual exercises, several prominent automation companies attended the course. Biosearch of Novato, CA, Genetic Design of Watertown, MA, were two of several companies to demonstrate automated synthesizers at the course. Biosearch presented their new SAM I synthesizer. The Genetic Design had developed their synthesizer from the design of its sister companies (Sequemat) solid phase peptide sequencer. The Genetic Design arranged with Dr Christian Birr (Max-Planck-Institute for Medical Research) a week before the event to convert his solid phase sequencer into the semi-automated synthesizer. The team led by Dr Alex Bonner and Rick Neves converted the unit and transported it to Darmstadt for the event and installed into the Biochemistry lab at the Technische Hochschule. As the system was semi-automatic, the user injected the next base to be added to the growing sequence during each cycle. The system worked well and produced a series of test tubes filled with bright red trityl color indicating complete coupling at each step. This system was later fully automated by inclusion of an auto injector and was designated the Model 25A.

Mid to Large Scale Oligonucleotide Synthesis

Large scale oligonucleotide synthesizers were often developed by augmenting the capabilities of a preexisting instrument platform. One of the first mid scale synthesizers appeared in the late 1980s, manufactured by the Biosearch company in Novato, CA (The 8800). This platform was originally designed as a peptide synthesizer and made use of a fluidized bed reactor essential for accommodating the swelling characteristics of polystyrene supports used in the Merrifield methodology. Oligonucleotide synthesis involved the use of CPG (controlled pore glass) which is a rigid support and is more suited for column reactors as described above. The scale of the 8800 was limited to the flow rate required to fluidize the support. Some novel reactor designs as well as higher than normal pressures enabled the 8800 to achieve scales that would prepare 1 mmole of oligonucleotide. In the mid 1990s several companies developed platforms that were based on semi-preparative and preparative liquid chromatographs. These systems were well suited for a column reactor approach. In most cases all that was required was to augment the number of fluids that could be delivered to the column. Oligo synthesis requires a minimum of 10 and liquid chromatographs usually accommodate 4. This was an easy design task and some semi-automatic strategies worked without any modifications to the preexisting LC equipment. PerSeptive Biosystems as well as Pharmacia (GE) were two of several companies that developed synthesizers out of liquid chromatographs. Genomic Technologies, Inc. was one of the few companies to develop a large scale oligonucleotide synthesizer that was, from the ground up, an oligonucleotide synthesizer. The initial platform called the VLSS for very large scale synthesizer utilized large Pharmacia liquid chromatograph columns as reactors and could synthesize up to 75 millimoles of material. Many oligonucleotide synthesis factories designed and manufactured their own custom platforms and little is known due to the designs being proprietary. The VLSS design continued to be refined and is continued in the QMaster synthesizer which is a scaled down platform providing milligram to gram amounts of synthetic oligonucleotide.

The current practices of synthesis of chemically modified oligonucleotides on large scale have been recently reviewed.

Synthesis of Oligonucleotide Microarrays

One may visualize an oligonucleotide microarray as a miniature multi-well plate where physical dividers between the wells (plastic walls) are intentionally removed. With respect to the chemistry, synthesis of oligonucleotide microarrays is different from the conventional oligonucleotide synthesis in two respects:

5'-*O*-MeNPOC-protected nucleoside phosphoramidite.

- Oligonucleotides remain permanently attached to the solid phase, which requires the use of linkers that are stable under the conditions of the final deprotection procedure.

- The absence of physical dividers between the sites occupied by individual oligonucleotides, a very limited space on the surface of the microarray (one oligonucleotide sequence occupies a square 25×25 μm) and the requirement of high fidelity of oligonucleotide synthesis dictate the use of site-selective 5'-deprotection techniques. In one approach, the removal of the 5'-O-DMT group is effected by electrochemical generation of the acid at the required site(s). Another approach uses 5'-O-(α-methyl-6-nitropiperonyloxycarbonyl) (MeNPOC) protecting group, which can be removed by irradiation with UV light of 365 nm wavelength.

Post-synthetic Processing

After the completion of the chain assembly, the solid support-bound oligonucleotide is fully protected:

- The 5'-terminal 5'-hydroxy group is protected with DMT group;

- The internucleosidic phosphate or phosphorothioate moieties are protected with 2-cyanoethyl groups;

- The exocyclic amino groups in all nucleic bases except for T and U are protected with acyl protecting groups.

To furnish a functional oligonucleotide, all the protecting groups have to be removed. The N-acyl base protection and the 2-cyanoethyl phosphate protection may be, and is often removed simultaneously by treatment with inorganic bases or amines. However, the applicability of this method is limited by the fact that the cleavage of 2-cyanoethyl phosphate protection gives rise to acrylonitrile as a side product. Under the strong basic conditions required for the removal of N-acyl protection, acrylonitrile is capable of alkylation of nucleic bases, primarily, at the N3-position of thymine and uracil residues to give the respective N3-(2-cyanoethyl) adducts via Michael reaction. The formation of these side products may be avoided by treating the solid support-bound oligonucleotides with solutions of bases in an organic solvent, for instance, with 50% triethylamine in acetonitrile or 10% diethylamine in acetonitrile. This treatment is strongly recommended for medium- and large scale preparations and is optional for syntheses on small scale where the concentration of acrylonitrile generated in the deprotection mixture is low.

Regardless of whether the phosphate protecting groups were removed first, the solid support-bound oligonucleotides are deprotected using one of the two general approaches.

- Most often, 5'-DMT group is removed at the end of the oligonucleotide chain assembly. The oligonucleotides are then released from the solid phase and deprotected (base and phosphate) by treatment with aqueous ammonium hydroxide, aqueous methylamine, their mixtures, gaseous ammonia or methylamine or, less commonly, solutions of other primary amines or alkalies at ambient or elevated temperature. This removes all remaining protection groups from 2'-deoxyoligonucleotides, resulting in a reaction mixture containing the desired product. If the oligonucleotide contains any 2'-O-protected ribonucleotide residues, the deprotection protocol includes the second step where the 2'-O-protecting silyl groups are removed by treatment with fluoride ion by various methods. The fully deprotected product is used as is, or the desired oligonucleotide can be purified by a number of methods. Most commonly, the crude product is desalted using ethanol precipitation, size exclusion chromatography, or reverse-phase HPLC. To eliminate unwanted truncation products, the oligonucleotides can be purified via polyacrylamide gel electrophoresis or anion-exchange HPLC followed by desalting.

- The second approach is only used when the intended method of purification is reverse-phase

HPLC. In this case, the 5'-terminal DMT group that serves as a hydrophobic handle for purification is kept on at the end of the synthesis. The oligonucleotide is deprotected under basic conditions as described above and, upon evaporation, is purified by reverse-phase HPLC. The collected material is then detritylated under aqueous acidic conditions. On small scale (less than 0.01–0.02 mmol), the treatment with 80% aqueous acetic acid for 15–30 min at room temperature is often used followed by evaporation of the reaction mixture to dryness in vacuo.

- For some applications, additional reporter groups may be attached to an oligonucleotide using a variety of post-synthetic procedures.

Characterization

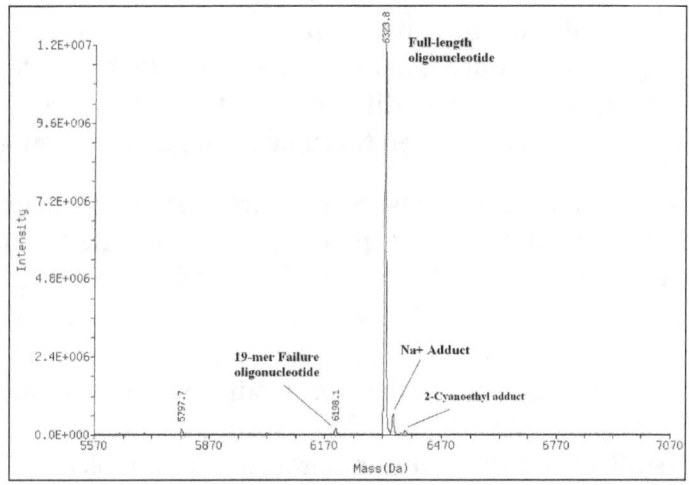

Deconvoluted ES MS of crude oligonucleotide 5'-DMT-T$_{20}$
(calculated mass 6324.26 Da).

As with any other organic compound, it is prudent to characterize synthetic oligonucleotides upon their preparation. In more complex cases (research and large scale syntheses) oligonucleotides are characterized after their deprotection and after purification. Although the ultimate approach to the characterization is sequencing, a relatively inexpensive and routine procedure, the considerations of the cost reduction preclude its use in routine manufacturing of oligonucleotides. In day-by-day practice, it is sufficient to obtain the molecular mass of an oligonucleotide by recording its mass spectrum. Two methods are currently widely used for characterization of oligonucleotides: electrospray mass spectrometry (ES MS) and matrix-assisted laser desorption/ionization time-of-flight mass spectrometry (MALDI-TOF). To obtain informative spectra, it is very important to exchange all metal ions that might be present in the sample for ammonium or trialkylammonium [e.c. triethylammonium, $(C_2H_5)_3NH^+$] ions prior to submitting a sample to the analysis by either of the methods.

- In ES MS spectrum, a given oligonucleotide generates a set of ions that correspond to different ionization states of the compound. Thus, the oligonucleotide with molecular mass M generates ions with masses $(M - nH)/n$ where M is the molecular mass of the oligonucleotide in the form of a free acid (all negative charges of internucleosidic phosphodiester groups are neutralized with H^+), n is the ionization state, and H is the atomic mass of hydrogen (1 Da). Most useful for characterization are the ions with n ranging from 2 to 5.

Software supplied with the more recently manufactured instruments is capable of performing a deconvolution procedure that is, it finds peaks of ions that belong to the same set and derives the molecular mass of the oligonucleotide.

- To obtain more detailed information on the impurity profile of oligonucleotides, liquid chromatography-mass spectrometry (LC-MS or HPLC-MS) or capillary electrophoresis mass spectrometry (CEMS) are used.

POST-TRANSCRIPTIONAL MODIFICATION

Post-transcriptional modification or co-transcriptional modification is a set of biological processes common to most eukaryotic cells by which an RNA primary transcript is chemically altered following transcription from a gene to produce a mature, functional RNA molecule that can then leave the nucleus and perform any of a variety of different functions in the cell. There are many types of post-transcriptional modifications achieved through a diverse class of molecular mechanisms.

Perhaps the most notable example is the conversion of precursor messenger RNA transcripts into mature messenger RNA that is subsequently capable of being translated into protein. This process includes three major steps that significantly modify the chemical structure of the RNA molecule: the addition of a 5' cap, the addition of a 3' polyadenylated tail, and RNA splicing. Such processing is vital for the correct translation of eukaryotic genomes because the initial precursor mRNA produced by transcription often contains both exons (coding sequences) and introns (non-coding sequences); splicing removes the introns and links the exons directly, while the cap and tail facilitate the transport of the mRNA to a ribosome and protect it from molecular degradation.

Post-transcriptional modifications may also occur during the processing of other transcripts which ultimately become transfer RNA, ribosomal RNA, or any of the other types of RNA used by the cell.

mRNA Processing

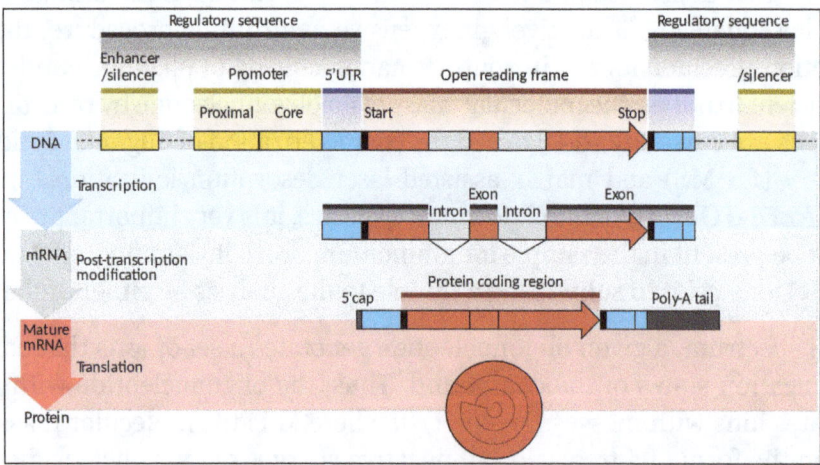

The above image shows: The structure of a prokaryotic operon of protein-coding genes. Regulatory sequence controls when expression occurs for the multiple protein coding regions (red). Promoter, operator and enhancer regions (yellow) regulate the transcription of the gene into an mRNA. The mRNA untranslated regions (blue) regulate translation into the final protein products.

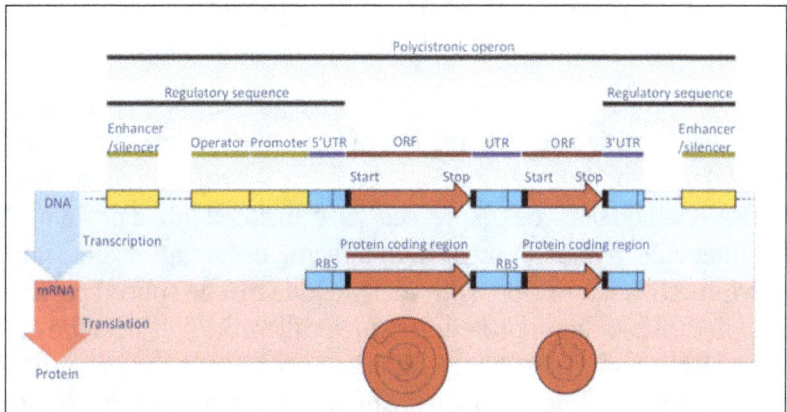

The pre-mRNA molecule undergoes three main modifications. These modifications are 5' capping, 3' polyadenylation, and RNA splicing, which occur in the cell nucleus before the RNA is translated.

5' Processing

Capping

Capping of the pre-mRNA involves the addition of 7-methylguanosine (m⁷G) to the 5' end. To achieve this, the terminal 5' phosphate requires removal, which is done with the aid of a phosphatase enzyme. The enzyme guanosyl transferase then catalyses the reaction, which produces the diphosphate 5' end. The diphosphate 5' end then attacks the alpha phosphorus atom of a GTP molecule in order to add the guanine residue in a 5'5' triphosphate link. The enzyme (guanine-N^7-)-methyltransferase ("cap MTase") transfers a methyl group from S-adenosyl methionine to the guanine ring. This type of cap, with just the (m⁷G) in position is called a cap 0 structure. The ribose of the adjacent nucleotide may also be methylated to give a cap 1. Methylation of nucleotides downstream of the RNA molecule produce cap 2, cap 3 structures and so on. In these cases the methyl groups are added to the 2' OH groups of the ribose sugar. The cap protects the 5' end of the primary RNA transcript from attack by ribonucleases that have specificity to the 3'5' phosphodiester bonds.

3' Processing

Cleavage and Polyadenylation

The pre-mRNA processing at the 3' end of the RNA molecule involves cleavage of its 3' end and then the addition of about 250 adenine residues to form a poly(A) tail. The cleavage and adenylation reactions occur if a polyadenylation signal sequence (5'- AAUAAA-3') is located near the 3' end of the pre-mRNA molecule, which is followed by another sequence, which is usually (5'-CA-3') and is the site of cleavage. A GU-rich sequence is also usually present further downstream on the pre-mRNA molecule. After the synthesis of the sequence elements, two multisubunit proteins called cleavage and polyadenylation specificity factor (CPSF) and cleavage stimulation factor (CStF) are transferred from RNA Polymerase II to the RNA molecule. The two factors bind to the sequence elements. A protein complex forms and contains additional cleavage factors and the enzyme Polyadenylate Polymerase (PAP). This complex cleaves the RNA between the polyadenylation sequence and the GU-rich sequence at the cleavage site marked by the (5'-CA-3') sequences.

Poly(A) polymerase then adds about 200 adenine units to the new 3' end of the RNA molecule using ATP as a precursor. As the poly(A) tail is synthesised, it binds multiple copies of poly(A) binding protein, which protects the 3'end from ribonuclease digestion.

Splicing

RNA splicing is the process by which introns, regions of RNA that do not code for proteins, are removed from the pre-mRNA and the remaining exons connected to re-form a single continuous molecule. Exons are sections of mRNA which become "expressed" or translated into a protein. They are the coding portions of a mRNA molecule. Although most RNA splicing occurs after the complete synthesis and end-capping of the pre-mRNA, transcripts with many exons can be spliced co-transcriptionally. The splicing reaction is catalyzed by a large protein complex called the spliceosome assembled from proteins and small nuclear RNA molecules that recognize splice sites in the pre-mRNA sequence. Many pre-mRNAs, including those encoding antibodies, can be spliced in multiple ways to produce different mature mRNAs that encode different protein sequences. This process is known as alternative splicing, and allows production of a large variety of proteins from a limited amount of DNA.

Histone mRNA Processing

Histones H2A, H2B, H3 and H4 form the core of a nucleosome and thus are called core histones. Processing of core histones is done differently because typical histone mRNA lacks several features of other eukaryotic mRNAs, such as poly(A) tail and introns. Thus, such mRNAs do not undergo splicing and their 3' processing is done independent of most cleavage and polyadenylation factors. Core histone mRNAs have a special stem-loop structure at 3-prime end that is recognized by a stem–loop binding protein and a downstream sequence, called histone downstream element (HDE) that recruits U7 snRNA. Cleavage and polyadenylation specificity factor 73 cuts mRNA between stem-loop and HDE

Histone variants, such as H2A.Z or H3.3, however, have introns and are processed as normal mRNAs including splicing and polyadenylation.

POST-TRANSLATIONAL MODIFICATION

Post-translational modification (PTM) refers to the covalent and generally enzymatic modification of proteins following protein biosynthesis. Proteins are synthesized by ribosomes translating mRNA into polypeptide chains, which may then undergo PTM to form the mature protein product. PTMs are important components in cell signaling, as for example when prohormones are converted to hormones.

Post-translational modifications can occur on the amino acid side chains or at the protein's C- or N- termini. They can extend the chemical repertoire of the 20 standard amino acids by modifying an existing functional group or introducing a new one such as phosphate. Phosphorylation is a very common mechanism for regulating the activity of enzymes and is the most common post-translational modification. Many eukaryotic proteins also have carbohydrate molecules attached to them in a process called glycosylation, which can promote protein folding and improve stability as well as serving regulatory functions. Attachment of lipid molecules, known as lipidation, often targets a protein or part of a protein attached to the cell membrane.

Other forms of post-translational modification consist of cleaving peptide bonds, as in processing a propeptide to a mature form or removing the initiator methionine residue. The formation of disulfide bonds from cysteine residues may also be referred to as a post-translational modification. For instance, the peptide hormone insulin is cut twice after disulfide bonds are formed, and a propeptide is removed from the middle of the chain; the resulting protein consists of two polypeptide chains connected by disulfide bonds.

Some types of post-translational modification are consequences of oxidative stress. Carbonylation is one example that targets the modified protein for degradation and can result in the formation of protein aggregates. Specific amino acid modifications can be used as biomarkers indicating oxidative damage.

Sites that often undergo post-translational modification are those that have a functional group that can serve as a nucleophile in the reaction: the hydroxyl groups of serine, threonine, and tyrosine; the amine forms of lysine, arginine, and histidine; the thiolate anion of cysteine; the carboxylates of aspartate and glutamate; and the N- and C-termini. In addition, although the amide of asparagine is a weak nucleophile, it can serve as an attachment point for glycans. Rarer modifications can occur at oxidized methionines and at some methylenes in side chains.

Post-translational modification of proteins can be experimentally detected by a variety of techniques, including mass spectrometry, Eastern blotting, and Western blotting.

PTMs Involving Addition of Functional Groups

Addition by an Enzyme in Vivo

Hydrophobic groups for membrane localization:

- Myristoylation (a type of acylation), attachment of myristate, a C14 saturated acid.

- Palmitoylation (a type of acylation), attachment of palmitate, a C16 saturated acid.

- Isoprenylation or prenylation, the addition of an isoprenoid group (e.g. farnesol and geranylgeraniol):

  ◦ Farnesylation.

  ◦ Geranilgeranilatyon.

  ◦ Glipyatyon, glycosylphosphatidylinositol (GPI) anchor formation via an amide bond to C-terminal tail.

Cofactors for enhanced enzymatic activity:

- Lipoylation (a type of acylation), attachment of a lipoate (C8) functional group.

- Flavin moiety (FMN or FAD) may be covalently attached.

- Heme C attachment via thioether bonds with cysteines.

- Phosphopantetheinylation, the addition of a 4'-phosphopantetheinyl moiety from coenzyme A, as in fatty acid, polyketide, non-ribosomal peptide and leucine biosynthesis.

- Retinylidene Schiff base formation.

Modifications of translation factors:

- Diphthamide formation (on a histidine found in eEF2).

- Ethanolamine phosphoglycerol attachment (on glutamate found in eEF1α).

- Hypusine formation (on conserved lysine of eIF5A (eukaryotic) and aIF5A (archaeal)).

- Beta-Lysine addition on a conserved lysine of the elongation factor P (EFP) in most bacteria. EFP is an homolog to eIF5A (eukaryotic) and aIF5A (archaeal).

Smaller chemical groups:

- Acylation, e.g. *O*-acylation (esters), *N*-acylation (amides), *S*-acylation (thioesters):

  ◦ Acetylation, the addition of an acetyl group, either at the N-terminus of the protein or at lysine residues. The reverse is called deacetylation.

  ◦ Formylation.

- Alkylation, the addition of an alkyl group, e.g. methyl, ethyl:

  ◦ methylation the addition of a methyl group, usually at lysine or arginine residues. The reverse is called demethylation.

- Amidation at C-terminus. Formed by oxidative dissociation of a C-terminal Gly residue.

- Amide bond formation:

  ◦ Amino acid addition:

    ▪ Arginylation, a tRNA-mediation addition.

    ▪ Polyglutamylation, covalent linkage of glutamic acid residues to the N-terminus of tubulin and some other proteins.

    ▪ Polyglycylation, covalent linkage of one to more than 40 glycine residues to the tubulin C-terminal tail.

- Butyrylation.

- Gamma-carboxylation dependent on Vitamin K.

- Glycosylation, the addition of a glycosyl group to either arginine, asparagine, cysteine, hydroxylysine, serine, threonine, tyrosine, or tryptophan resulting in a glycoprotein. Distinct from glycation, which is regarded as a nonenzymatic attachment of sugars.

- ◦ Polysialylation, addition of polysialic acid, PSA, to NCAM.

- Malonylation.

- Hydroxylation: Addition of an oxygen atom to the side-chain of a Pro or Lys residue.

- Iodination: Addition of an iodine atom to the aromatic ring of a tyrosine residue (e.g. in thyroglobulin).

- Nucleotide addition such as ADP-ribosylation.

- Phosphate ester (O-linked) or phosphoramidate (N-linked) formation:

  - ◦ Phosphorylation, the addition of a phosphate group, usually to serine, threonine, and tyrosine (O-linked), or histidine (N-linked).

  - ◦ Adenylylation, the addition of an adenylyl moiety, usually to tyrosine (O-linked), or histidine and lysine (N-linked).

  - ◦ Uridylylation, the addition of an uridylyl-group (i.e. uridine monophosphate, UMP), usually to tyrosine.

- Propionylation.

- Pyroglutamate formation.

- S-glutathionylation.

- S-nitrosylation.

- S-sulfenylation (aka S-sulphenylation), reversible covalent addition of one oxygen atom to the thiol group of a cysteine residue.

- S-sulfinylation, normally irreversible covalent addition of two oxygen atoms to the thiol group of a cysteine residue.

- S-sulfonylation, normally irreversible covalent addition of three oxygen atoms to the thiol group of a cysteine residue, resulting in the formation of a cysteic acid residue.

- Succinylation addition of a succinyl group to lysine.

- Sulfation, the addition of a sulfate group to a tyrosine.

Non-enzymatic Additions in Vivo

- Glycation, the addition of a sugar molecule to a protein without the controlling action of an enzyme.

- Carbamylation the addition of Isocyanic acid to a protein's N-terminus or the side-chain of Lys.

- Carbonylation the addition of carbon monoxide to other organic/inorganic compounds.

- Spontaneous isopeptide bond formation, as found in many surface proteins of Gram-positive bacteria.

Non-enzymatic Additions in Vitro

- Biotinylation: Covalent attachment of a biotin moiety using a biotinylation reagent, typically for the purpose of labeling a protein.

- Carbamylation: The addition of Isocyanic acid to a protein's N-terminus or the side-chain of Lys or Cys residues, typically resulting from exposure to urea solutions.

- Oxidation: Addition of one or more Oxygen atoms to a susceptible side-chain, principally of Met, Trp, His or Cys residues. Formation of disulfide bonds between Cys residues.

- Pegylation: Covalent attachment of polyethylene glycol (PEG) using a pegylation reagent, typically to the N-terminus or the side-chains of Lys residues. Pegylation is used to improve the efficacy of protein pharmaceuticals.

Other Proteins or Peptides

- ISGylation, the covalent linkage to the ISG15 protein (Interferon-Stimulated Gene 15).

- SUMOylation, the covalent linkage to the SUMO protein (Small Ubiquitin-related Modifier).

- Ubiquitination, the covalent linkage to the protein ubiquitin.

- Neddylation, the covalent linkage to Nedd.

- Pupylation, the covalent linkage to the prokaryotic ubiquitin-like protein.

Chemical Modification of Amino Acids

- Citrullination, or deimination, the conversion of arginine to citrulline.

- Deamidation, the conversion of glutamine to glutamic acid or asparagine to aspartic acid.

- Eliminylation, the conversion to an alkene by beta-elimination of phosphothreonine and phosphoserine, or dehydration of threonine and serine.

Structural Changes

- Disulfide bridges, the covalent linkage of two cysteine amino acids.

- Proteolytic cleavage, cleavage of a protein at a peptide bond.

- Isoaspartate formation, via the cyclisation of asparagine or aspartic acid amino-acid residues.

- Racemization.

- ◦ Of serine by protein-serine epimerase.

- ◦ Of alanine in dermorphin, a frog opioid peptide.

- ◦ Of methionine in deltorphin, also a frog opioid peptide.

- • Protein splicing, self-catalytic removal of inteins analogous to mRNA processing.

RECOMBINANT DNA

Recombinant DNA refers to the molecules of DNA from two different species that are inserted into a host organism to produce new genetic combinations that are of value to science, medicine, agriculture, and industry. Since the focus of all genetics is the gene, the fundamental goal of laboratory geneticists is to isolate, characterize, and manipulate genes. Although it is relatively easy to isolate a sample of DNA from a collection of cells, finding a specific gene within this DNA sample can be compared to finding a needle in a haystack. Consider the fact that each human cell contains approximately 2 metres (6 feet) of DNA. Therefore, a small tissue sample will contain many kilometres of DNA. However, recombinant DNA technology has made it possible to isolate one gene or any other segment of DNA, enabling researchers to determine its nucleotide sequence, study its transcripts, mutate it in highly specific ways, and reinsert the modified sequence into a living organism.

DNA Cloning

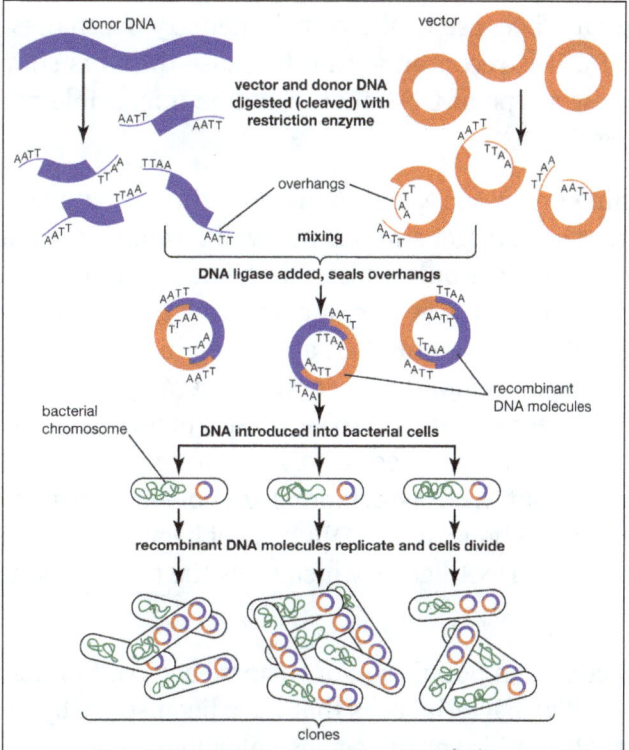

Steps involved in the engineering of a recombinant DNA molecule.

In biology a clone is a group of individual cells or organisms descended from one progenitor. This means that the members of a clone are genetically identical, because cell replication produces identical daughter cells each time. The use of the word clone has been extended to recombinant DNA technology, which has provided scientists with the ability to produce many copies of a single fragment of DNA, such as a gene, creating identical copies that constitute a DNA clone. In practice the procedure is carried out by inserting a DNA fragment into a small DNA molecule and then allowing this molecule to replicate inside a simple living cell such as a bacterium. The small replicating molecule is called a DNA vector (carrier). The most commonly used vectors are plasmids (circular DNA molecules that originated from bacteria), viruses, and yeast cells. Plasmids are not a part of the main cellular genome, but they can carry genes that provide the host cell with useful properties, such as drug resistance, mating ability, and toxin production. They are small enough to be conveniently manipulated experimentally, and, furthermore, they will carry extra DNA that is spliced into them.

Creating the Clone

The steps in cloning are as follows. DNA is extracted from the organism under study and is cut into small fragments of a size suitable for cloning. Most often this is achieved by cleaving the DNA with a restriction enzyme. Restriction enzymes are extracted from several different species and strains of bacteria, in which they act as defense mechanisms against viruses. They can be thought of as "molecular scissors," cutting the DNA at specific target sequences. The most useful restriction enzymes make staggered cuts; that is, they leave a single-stranded overhang at the site of cleavage. These overhangs are very useful in cloning because the unpaired nucleotides will pair with other overhangs made using the same restriction enzyme. So, if the donor DNA and the vector DNA are both cut with the same enzyme, there is a strong possibility that the donor fragments and the cut vector will splice together because of the complementary overhangs. The resulting molecule is called recombinant DNA. It is recombinant in the sense that it is composed of DNA from two different sources. Thus, it is a type of DNA that would be impossible naturally and is an artifact created by DNA technology.

The next step in the cloning process is to cut the vector with the same restriction enzyme used to cut the donor DNA. Vectors have target sites for many different restriction enzymes, but the most convenient ones are those that occur only once in the vector molecule. This is because the restriction enzyme then merely opens up the vector ring, creating a space for the insertion of the donor DNA segment. Cut vector DNA and donor DNA are mixed in a test tube, and the complementary ends of both types of DNA unite randomly. Of course, several types of unions are possible: donor fragment to donor fragment, vector fragment to vector fragment, and, most important, vector fragment to donor fragment, which can be selected for. Recombinant DNA associations form spontaneously in the above manner, but these associations are not stable because, although the ends are paired, the sugar-phosphate backbone of the DNA has not been sealed. This is accomplished by the application of an enzyme called DNA ligase, which seals the two segments, forming a continuous and stable double helix.

The mixture should now contain a population of vectors each containing a different donor insert. This solution is mixed with live bacterial cells that have been specially treated to make their cells more permeable to DNA. Recombinant molecules enter living cells in a process called transformation. Usually, only a single recombinant molecule will enter any individual bacterial cell. Once inside, the recombinant DNA molecule replicates like any other plasmid DNA molecule, and many copies are subsequently produced. Furthermore, when the bacterial cell divides, all of the daughter cells receive the recombinant plasmid, which again replicates in each daughter cell.

The original mixture of transformed bacterial cells is spread out on the surface of a growth medium in a flat dish (Petri dish) so that the cells are separated from one another. These individual cells are invisible to the naked eye, but as each cell undergoes successive rounds of cell division, visible colonies form. Each colony is a cell clone, but it is also a DNA clone because the recombinant vector has now been amplified by replication during every round of cell division. Thus, the Petri dish, which may contain many hundreds of distinct colonies, represents a large number of clones of different DNA fragments. This collection of clones is called a DNA library. By considering the size of the donor genome and the average size of the inserts in the recombinant DNA molecule, a researcher can calculate the number of clones needed to encompass the entire donor genome, or, in other words, the number of clones needed to constitute a genomic library.

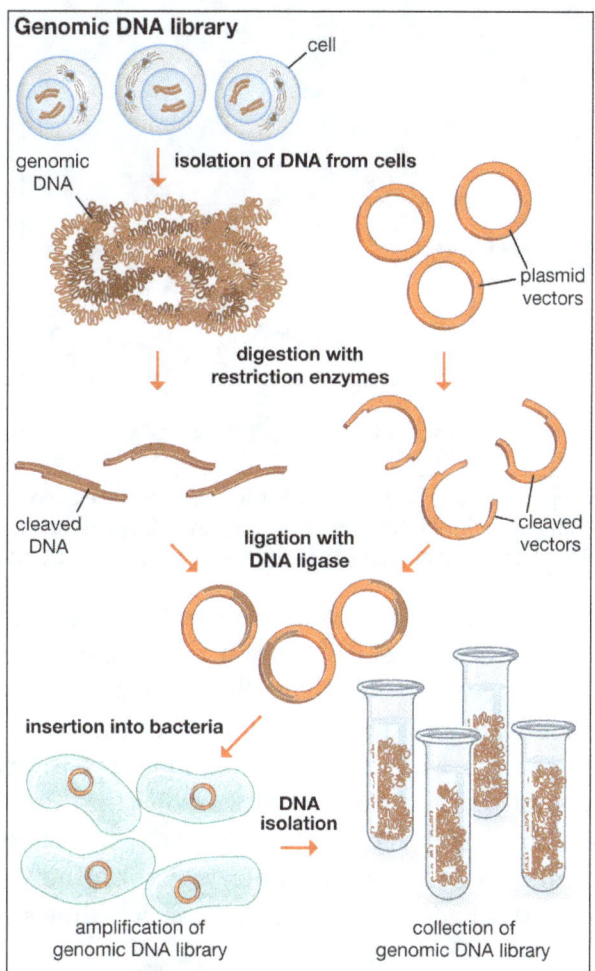

A genomic DNA library is a collection of DNA fragments that make up the full-length genome of an organism. A genomic library is created by isolating DNA from cells and then amplifying it using DNA cloning technology.

Another type of library is a cDNA library. Creation of a cDNA library begins with messenger ribonucleic acid (mRNA) instead of DNA. Messenger RNA carries encoded information from DNA to ribosomes for translation into protein. To create a cDNA library, these mRNA molecules are treated with the enzyme reverse transcriptase, which is used to make a DNA copy of an mRNA. The resulting DNA molecules are called complementary DNA (cDNA). A cDNA library represents a sampling of the transcribed genes, whereas a genomic library includes untranscribed regions.

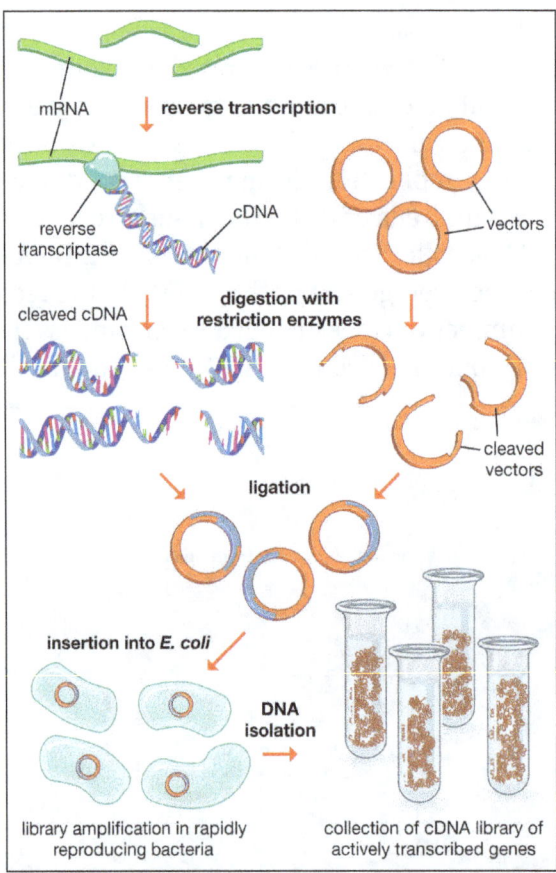

A cDNA library represents a collection of only the genes that are encoded into proteins by an organism. Complementary DNA, or cDNA, is created through reverse transcription of messenger RNA, and a library of cDNAs is generated using DNA cloning technology.

Both genomic and cDNA libraries are made without regard to obtaining functional cloned donor fragments. Genomic clones do not necessarily contain full-length copies of genes. Furthermore, genomic DNA from eukaryotes (cells or organisms that have a nucleus) contains introns, which are regions of DNA that are not translated into protein and cannot be processed by bacterial cells. This means that even full-sized genes are not translated in their entirety. In addition, eukaryotic regulatory signals are different from those used by prokaryotes (cells or organisms lacking internal membranes—i.e., bacteria). However, it is possible to produce expression libraries by slicing cDNA inserts immediately adjacent to a bacterial promoter region on the vector; in these expression libraries, eukaryotic proteins are made in bacterial cells, which allows several important technological applications.

Several bacterial viruses have also been used as vectors. The most commonly used is the lambda phage. The central part of the lambda genome is not essential for the virus to replicate in Escherichia coli, so this can be excised using an appropriate restriction enzyme, and inserts from donor DNA can be spliced into the gap. In fact, when the phage repackages DNA into its protein capsule, it includes only DNA fragments the same length of the normal phage genome.

Vectors are chosen depending on the total amount of DNA that must be included in a library. Cosmids are engineered vectors that are hybrids of plasmid and phage lambda; however, they can carry larger inserts than either pUC plasmids (plasmids engineered to produce a very high number of DNA copies but that can accommodate only small inserts) or lambda phage alone. Bacterial artificial chromosomes (BACs) are vectors based on F-factor (fertility factor) plasmids of E. coli and can carry much larger amounts of DNA. Yeast artificial chromosomes (YACs) are vectors based on autonomously replicating plasmids of Saccharomyces cerevisiae (baker's yeast). In yeast (a eukaryotic organism) a YAC behaves like a yeast chromosome and segregates properly into daughter cells. These vectors can carry the largest inserts of all and are used extensively in cloning large genomes such as the human genome.

Isolating the Clone

In general, cloning is undertaken in order to obtain the clone of one particular gene or DNA sequence of interest. The next step after cloning, therefore, is to find and isolate that clone among other members of the library. If the library encompasses the whole genome of an organism, then somewhere within that library will be the desired clone. There are several ways of finding it, depending on the specific gene concerned. Most commonly, a cloned DNA segment that shows homology to the sought gene is used as a probe. For example, if a mouse gene has already been cloned, then that clone can be used to find the equivalent human clone from a human genomic library. Bacterial colonies constituting a library are grown in a collection of Petri dishes. Then a porous membrane is laid over the surface of each plate, and cells adhere to the membrane. The cells are ruptured, and DNA is separated into single strands—all on the membrane. The probe is also separated into single strands and labeled, often with radioactive phosphorus. A solution of the radioactive probe is then used to bathe the membrane. The single-stranded probe DNA will adhere only to the DNA of the clone that contains the equivalent gene. The membrane is dried and placed against a sheet of radiation-sensitive film, and somewhere on the films a black spot will appear, announcing the presence and location of the desired clone. The clone can then be retrieved from the original Petri dishes.

DNA Sequencing

Once a segment of DNA has been cloned, its nucleotide sequence can be determined. The nucleotide sequence is the most fundamental level of knowledge of a gene or genome. It is the blueprint that contains the instructions for building an organism, and no understanding of genetic function or evolution could be complete without obtaining this information.

Uses

Knowledge of the sequence of a DNA segment has many uses, and some examples follow. First, it can be used to find genes, segments of DNA that code for a specific protein or phenotype. If a region of DNA has been sequenced, it can be screened for characteristic features of genes. For example, open reading frames (ORFs)—long sequences that begin with a start codon (three adjacent nucleotides; the sequence of a codon dictates amino acid production) and are uninterrupted by stop codons (except for one at their termination)—suggest a protein-coding region. Also, human genes are generally adjacent to so-called CpG islands—clusters of cytosine and guanine, two of the nucleotides that make up DNA. If a gene with a known phenotype (such as a disease gene in humans) is known to be in the chromosomal region sequenced, then unassigned genes in the region will become candidates for that function. Second, homologous DNA sequences of different organisms can be compared in order to plot evolutionary relationships both within and between species. Third, a gene sequence can be screened for functional regions. In order to determine the function of a gene, various domains can be identified that are common to proteins of similar function. For example, certain amino acid sequences within a gene are always found in proteins that span a cell membrane; such amino acid stretches are called transmembrane domains. If a transmembrane domain is found in a gene of unknown function, it suggests that the encoded protein is located in the cellular membrane. Other domains characterize DNA-binding proteins. Several public databases of DNA sequences are available for analysis by any interested individual.

Methods

The two basic sequencing approaches are the Maxam-Gilbert method, discovered by and named for American molecular biologists Allan M. Maxam and Walter Gilbert, and the Sanger method, discovered by English biochemist Frederick Sanger. In the most commonly used method, the Sanger method, DNA chains are synthesized on a template strand, but chain growth is stopped when one of four possible dideoxy nucleotides, which lack a 3′ hydroxyl group, is incorporated, thereby preventing the addition of another nucleotide. A population of nested, truncated DNA molecules results that represents each of the sites of that particular nucleotide in the template DNA. These molecules are separated in a procedure called electrophoresis, and the inferred nucleotide sequence is deduced using a computer.

In Vitro Mutagenesis

Another use of cloned DNA is in vitro mutagenesis in which a mutation is produced in a segment of cloned DNA. The DNA is then inserted into a cell or organism, and the effects of the mutation are studied. Mutations are useful to geneticists in enabling them to investigate the components of any biological process. However, traditional mutational analysis relied on the occurrence of random spontaneous mutations—a hit-or-miss method in which it was impossible to predict the precise type or position of the mutations obtained. In vitro mutagenesis, however, allows specific mutations to be tailored for type and for position within the gene. A cloned gene is treated in the test tube (in vitro) to obtain the specific mutation desired, and then this fragment is reintroduced into the living cell, where it replaces the resident gene.

One method of in vitro mutagenesis is oligonucleotide-directed mutagenesis. A specific point in a sequenced gene is pinpointed for mutation. An oligonucleotide, a short stretch of synthetic DNA of the desired sequence, is made chemically. For example, the oligonucleotide might have adenine in one specific location instead of guanine. This oligonucleotide is hybridized to the complementary strand of the cloned gene; it will hybridize despite the one base pair mismatch. Various enzymes are added to allow the oligonucleotide to prime the synthesis of a complete strand within the vector. When the vector is introduced into a bacterial cell and replicates, the mutated strand will act as a template for a complementary strand that will also be mutant, and thus a fully mutant molecule is obtained. This fully mutant cloned molecule is then reintroduced into the donor organism, and the mutant DNA replaces the resident gene.

Another version of in vitro mutagenesis is gene disruption, or gene knockout. Here, the resident functional gene is replaced by a completely nonfunctional copy. The advantage of this technique over random mutagenesis is that specific genes can be knocked out at will, leaving all other genes untouched by the mutagenic procedure.

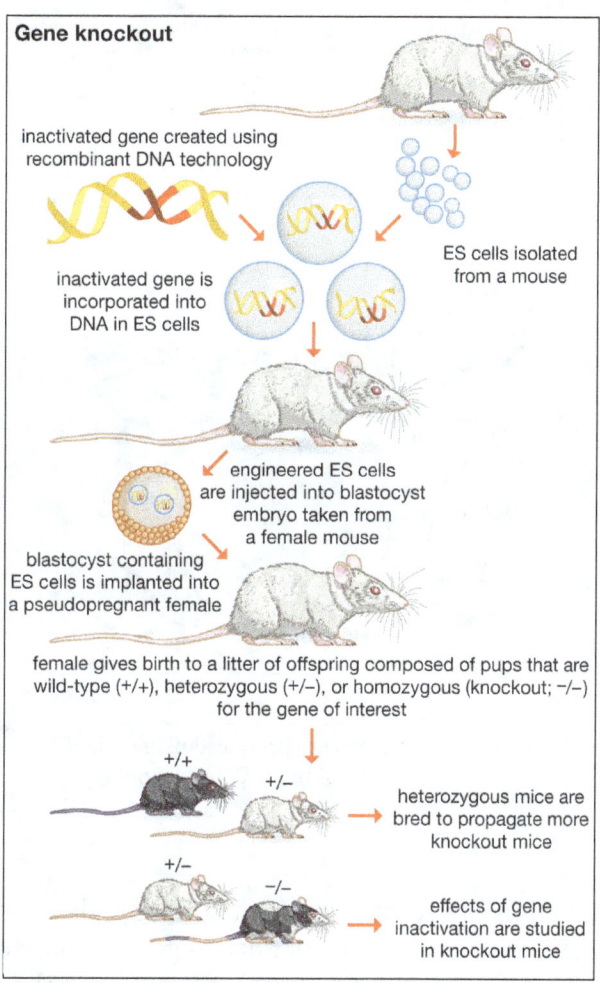

In gene knockout a functional gene is replaced by an inactivated gene that is created using recombinant DNA technology. When a gene is "knocked out," the resulting mutant phenotype (observable characteristics) often reveals the gene's biological function.

Genetically Modified Organisms

The ability to obtain specific DNA clones using recombinant DNA technology has made it possible to add the DNA of one organism to the genome of another. The added gene is called a transgene. The transgene inserts itself into a chromosome and is passed to the progeny as a new component of the genome. The resulting organism carrying the transgene is called a transgenic organism or a genetically modified organism (GMO). In this way, a "designer organism" is made that contains some specific change required for an experiment in basic genetics or for improvement of some commercial strain. Several transgenic plants have been produced. Genes for toxins that kill insects have been introduced in several species, including corn and cotton. Bacterial genes that confer resistance to herbicides also have been introduced into crop plants. Other plant transgenes aim at improving the nutritional value of the plant.

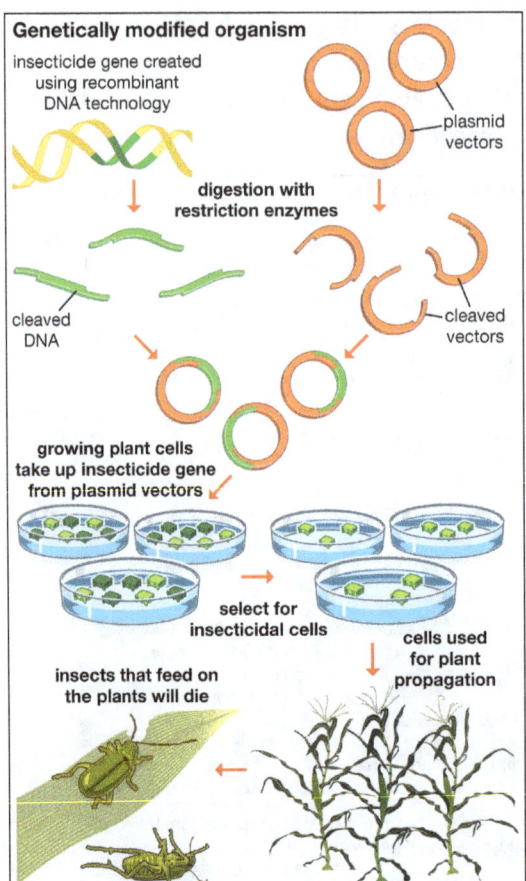

Genetically modified organisms are produced using scientific methods
that include recombinant DNA technology.

Gene Therapy

Gene therapy is the introduction of a normal gene into an individual's genome in order to repair a mutation that causes a genetic disease. When a normal gene is inserted into a mutant nucleus, it most likely will integrate into a chromosomal site different from the defective allele; although this may repair the mutation, a new mutation may result if the normal gene integrates into another functional gene. If the normal gene replaces the mutant allele, there is a chance that the transformed cells will proliferate and produce enough normal gene product for the entire body to be restored to the undiseased phenotype. So far, human gene therapy has been attempted only on somatic (body) cells for diseases such as cancer and severe combined immunodeficiency syndrome (SCIDS). Somatic cells cured by gene therapy may reverse the symptoms of disease in the treated individual, but the modification is not passed on to the next generation. Germinal gene therapy aims to place corrected cells inside the germ line (e.g., cells of the ovary or testis). If this is achieved, these cells will undergo meiosis and provide a normal gametic contribution to the next generation. Germinal gene therapy has been achieved experimentally in animals but not in humans.

Reverse Genetics

Recombinant DNA technology has made possible a type of genetics called reverse genetics. Traditionally, genetic research starts with a mutant phenotype, and by Mendelian crossing analysis, a researcher is able to attribute the phenotype to a specific gene. Reverse genetics travels in precisely the opposite direction. Researchers begin with a gene of unknown function and use molecular analysis to determine its phenotype. One important tool in reverse genetics is gene knockout. By mutating the cloned gene of unknown function and using it to replace the resident copy or copies, the resultant mutant phenotype will show which biological function this gene normally controls.

Diagnostics

Recombinant DNA technology has led to powerful diagnostic procedures useful in both medicine and forensics. In medicine these diagnostic procedures are used in counseling prospective parents as to the likelihood of having a child with a particular disease, and they are also used in the prenatal prediction of genetic disease in the fetus. Researchers look for specific DNA fragments that are located in close proximity to the gene that causes the disease of concern. These fragments, called restriction fragment length polymorphisms (RFLPs), often serve as effective "genetic markers." In forensics, DNA fragments called variable number tandem repeats (VNTRs), which are highly variable between individuals, are employed to produce what is called a "DNA fingerprint." A DNA fingerprint can be used to determine if blood or other body fluids left at the scene of a crime belongs to a suspect.

Genomics

The genetic analysis of entire genomes is called genomics. Such a broadscale analysis has been made possible by the development of recombinant DNA technology. In humans, knowledge of the entire genome sequence has facilitated searching for genes that produce hereditary diseases. It is also capable of revealing a set of proteins—produced at specific times, in specific tissues, or in specific diseases—that might be targets for therapeutic drugs. Genomics also allows the comparison of one genome with another, leading to insights into possible evolutionary relationships between organisms.

Genomics has two subdivisions: structural genomics and functional genomics. Structural genomics is based on the complete nucleotide sequence of a genome. Each member of a library of clones is physically manipulated by robots and sequenced by automatic sequencing machines, enabling a very high throughput of DNA. The resulting sequences are then assembled by a computer into a complete sequence for every chromosome. The complete DNA sequence is scanned by computer to find the positions of open reading frames (ORFs), or prospective genes. The sequences are then compared to the sequences of known genes from other organisms, and possible functions are assigned. Some ORFs remain unassigned, awaiting further research.

Functional genomics attempts to understand function at the broadest level (the genomic level). In one approach, gene functions of as many ORFs as possible are assigned in an attempt to obtain a full set of proteins encoded by the genome (called a proteome). The proteome broadly defines all the cellular functions used by the organism. Function in relation to specific developmental stages also is assessed by trying to identify the "transcriptome," the set of mRNA transcripts made at specific developmental stages. The practical approach utilizes microarrays—glass plates the size of a microscope slide imprinted with tens of thousands of ordered DNA samples, each representing one gene (either a clone or a synthesized segment). The mRNA preparation under test is labeled with a fluorescent dye, and the microarray is bathed in this mRNA. Fluorescent spots appear on the array indicating which mRNAs were present, thus defining the transcriptome.

Protein Manufacture

Recombinant DNA procedures have been used to convert bacteria into "factories" for the synthesis of foreign proteins. This technique is useful not only for preparing large amounts of protein for basic research but also for producing valuable proteins for medical use. For example, the genes for human proteins such as growth hormone, insulin, and blood-clotting factor can be commercially manufactured. Another approach to producing proteins via recombinant DNA technology is to introduce the desired gene into the genome of an animal, engineered in such a way that the protein is secreted in the animal's milk, facilitating harvesting.

References

- Dna-replication-3981005: thoughtco.com, Retrieved 9 May, 2019

- Samson RY, Bell SD (2014). "Archaeal chromosome biology". Journal of Molecular Microbiology and Biotechnology. 24 (5–6): 420–27. Doi:10.1159/000368854. PMC 5175462. PMID 25732343.

- Molecular-mechanism-of-dna-replication, dna-replication, dna-as-the-genetic-material, science: khanacademy.org, Retrieved 20 June, 2019

- Mankertz P (2008). "Molecular Biology of Porcine Circoviruses". Animal Viruses: Molecular Biology. Caister Academic Press. ISBN 978-1-904455-22-6

- Recombinant-DNA-technology, science: britannica.com, Retrieved 21 July, 2019

INDEX

A

Adenine, 9, 39, 51, 53, 64, 67, 69, 77, 94, 102, 105-106, 194, 207-208, 218

Adenosine Diphosphate, 6, 76

Adenosine Triphosphate, 5, 18, 51, 76

Amino Acids, 4, 6, 16, 21, 33, 36-44, 48, 50, 57, 60, 64-65, 84, 86-88, 94, 164, 189-190, 192, 208, 212

Antibiotic Resistance, 128, 150, 152

B

Blot, 83, 103-105

Bone Morphogenetic Proteins, 31

Buoyant Density, 55

C

Calcium Phosphate, 13, 173-174

Capillary Electrophoresis, 107, 121-124, 159, 206

Cell Culture, 163-166, 168-171, 174, 176

Cell Differentiation, 27-30, 32

Cell Nucleus, 8-10, 28-29, 40, 64, 191, 207

Cell Signaling, 35, 38, 44, 208

Cellulose Fibrils, 19

Chain-termination Sequencing, 117, 120

Cytokinesis, 23-24

Cytosine, 9, 30, 51, 53, 67, 94, 105-106, 115, 194, 217

D

Deoxyribonucleic Acid, 1, 4, 36, 51-52

Deoxyribonucleotide Triphosphate, 125-126

Dna Double Helix, 8, 22, 30, 53, 67, 74, 179, 181-182, 190

Dna Fingerprinting, 166

Dna Replication, 8, 22-24, 30, 37, 44, 62-63, 66-79, 95, 119, 178-182

Dot Blot, 105

Dye-terminator Sequencing, 121, 123

E

Embryonic Cells, 15, 23, 25, 31

Endoplasmic Reticulum, 2, 10, 46, 85-86, 190

Escherichia Coli, 36, 43, 107, 216

Eukaryotic Cells, 3, 34-35, 39, 64, 74-76, 84, 87, 173, 178, 188, 191, 206

Extracellular Matrix, 12-18, 164-165, 170

F

Fibrous Proteins, 12, 42, 45

G

Galacturonic Acid, 19-21

Gel Electrophoresis, 46, 91, 94-95, 97-98, 104-105, 117, 120, 189, 204

Gene Expression, 29, 32, 60, 65, 77, 84, 110, 119, 140, 142-143, 145, 147-148, 156, 158, 160-161, 163, 169, 185

Gene Knockout, 219, 221

Gene Therapy, 154, 173, 177, 220

Genetic Code, 9, 37, 39-40, 60, 65, 85, 87-88, 90, 192

Genetically Modified Organism, 219

Glycolysis, 6-7

Golgi Apparatus, 2, 10-11, 19

Green Fluorescent Protein, 47, 74

Guanine, 9, 30, 39, 51, 53, 64, 67, 94, 105-106, 207, 217-218

I

Immunohistochemistry, 28, 38, 47

In Vitro Mutagenesis, 218-219

Ion Semiconductor Sequencing, 125, 127

L

Lagging Strand, 70-72, 74, 181-182

M

Mass Spectrometry, 38, 48, 117, 189, 205-206, 209

Meiosis, 24-25, 178, 187, 220

Messenger Rna, 9, 33, 40, 56, 60, 64, 78, 84, 142, 189, 191-192, 206, 215-216

Mitochondria, 2, 8, 10-11, 34, 47, 52, 58, 184

Mitosis, 22-25, 176, 178

Molecular Cloning, 91, 94, 98, 149-150

Myeloma, 168, 172

N

Nanopore Sequencing, 115, 128-130, 132

Next Generation Sequencing, 121, 134, 137, 145, 177

Nucleic Acids, 1, 16, 36-38, 46, 51, 62, 77, 104, 118, 130, 137, 150, 173, 175, 193, 200

Nucleotide Sequence, 9, 33, 37, 39, 57, 60, 65-66, 94, 114, 154, 182, 213, 217-218, 221

O

Oligonucleotide Synthesis, 178, 193-194, 196-203

Oligosaccharins, 21

Oxidative Phosphorylation, 7, 10

P

Peptide Bonds, 37, 39, 209

Polymerase Chain Reaction, 67, 91, 95, 108, 123, 150

Polysaccharide, 12-13, 37, 91

Post-transcriptional Modification, 40, 191, 206

Prokaryotic Cells, 34, 64, 87, 191

Protein Synthesis, 33-34, 40, 51, 56-57, 60, 87-88, 189-190

R

Radioactive Phosphorus, 121, 217

Replication Foci, 76

Reverse Transcription, 62, 64-65, 83-84, 118, 140-143, 145, 156, 159, 216

Ribonucleic Acid, 1, 4, 37, 51, 215

Ribozymes, 4, 9, 57, 78

Rna Polymerase, 30, 40, 55-56, 60, 64, 77-81, 118, 191, 207

S

Sanger Sequencing, 107, 110, 117-119, 122-124, 128, 135-137

Sequence Motif, 43

T

Taq Polymerase, 95-97, 102

Thermal Cycling, 122-124

Thymine, 9, 51, 53, 67, 69, 78-79, 94, 105-106, 191, 194, 204

Transfer Rna, 33, 40, 56-57, 60, 70, 78, 85, 87, 189, 192, 206

Tricarboxylic Acid, 6-7, 10

U

Uracil, 9, 39, 51, 64, 78-79, 191, 194, 204

V

Viral Genomes, 57, 183

W

Western Blot, 104-105

X

X-ray Crystallography, 38, 48-49

Z

Zero-mode Waveguide, 135-136